CHASING
THE SUN

ALSO BY LINDA GEDDES

*Bumpology: The Myth-Busting Pregnancy Book
for Curious Parents-to-Be*

CHASING THE SUN

How the Science of Sunlight Shapes Our Bodies and Minds

LINDA GEDDES

PEGASUS BOOKS
NEW YORK LONDON

CHASING THE SUN

Pegasus Books Ltd.
148 West 37th Street, 13th Floor
New York, NY 10018

First Pegasus Books hardcover edition October 2019

ISBN: 978-1-64313-217-4

10 9 8 7 6 5 4 3 2 1

Printed in the United States of America
Distributed by W. W. Norton & Company, Inc.

For my mum,
who tracks the lengthening evenings
with brightening spirits

Contents

Introduction

IF EVER YOU NEED a reminder of the sun's awesome power, the Mojave Desert is a good place to start. In summer, when the daytime temperature frequently hits 49°C (120°F), stepping outside feels like opening the door to a giant furnace.

The local flora and fauna arm themselves against this heat: Joshua trees grow tough, concave spikes for leaves, minimising water loss and channelling what little rain does fall down towards their trunk and roots; jackrabbits develop enormous ears with shallow blood vessels that allow their body heat to quickly evaporate. Other creatures are nocturnal or emerge only at dawn or dusk to avoid the sun's heat; while yet others, such as the desert tortoise, sleep out the entire summer in underground burrows. Then there are vultures, which cool off by peeing onto their own legs.

Humans are less well equipped for such harsh conditions. In the Sonora Desert immediately to the south, hundreds of Central American migrants meet their deaths each year while trying to cross the border into the USA, the sun stripping away their body fluids and causing them to overheat.

Yet the sun's power also creates opportunities. Plants harness its rays to generate food, while shimmering solar farms are springing up, bent on transforming those rays into electricity. At the largest of these – the Ivanpah Solar Plant, 45 miles south-west of Las Vegas – a glittering sea of sun-tracking mirrors captures and focuses sunlight onto three boiler-topped towers, which drive turbines that supply electricity to hundreds of thousands of homes. Pity any bird crossing the path of these concentrated sunbeams, though: they're known as 'streamers', referencing the wisps of white smoke left in their wake as they are instantly incinerated. Across the ages, in civilisations separated by thousands of miles of land or ocean, people have revered the sun as both a creator and a destroyer – a relationship that continues to this day.

In Las Vegas, though, which rises out of this hostile landscape in defiance, the sun has been dethroned. At night, the neon-soaked Strip is reportedly the brightest place on earth, while the strongest artificial light on the planet emits from the tip of the Luxor Resort and Casino's glass and steel pyramid: each night, it sends this powerful 'sky beam' upwards, as if issuing a direct challenge to our nearest star. On a clear night, airline passengers can see the beam from some 275 miles away, and their pilots use it to help them navigate. The artificial light also confuses the navigation systems of insects, luring them to their deaths: the concentrated swarms provide a buffet for bats, which are in turn feasted upon by swooping owls.

Elsewhere in Vegas, realising the power it wields over our minds and spirits, resort owners have deliberately banished the sun from their casino floors. The 24-hour cycle of light and darkness is crucial to our internal sense of time; if there are no windows, it's easier for gamblers to lose track of it and stay for hours longer than they mean to – particularly if artificial light is deployed to wake them up. Some casinos even go so far as forbidding dealers from wearing a watch, so that if anyone asks the time, they can't say. Chairs are ergonomically designed to allow players to sit for hours in comfort, while oxygen is pumped in to boost alertness.

In this twilit world, artificial light reigns supreme, and it can have a huge effect on us: strategically placed spotlights draw consumers towards jangling, flashing slot machines; but the colour of the lighting can also be deliberately tweaked to manipulate people's behaviour. Blue-white light simulates daylight and makes people feel more alert, which can fool them into lingering longer at tables and slot machines. Meanwhile, red light can raise our level of physiological arousal: a study has shown that people gamble more money, place more bets and choose riskier options under red light, compared to blue. Pair that red light with fast music, another study showed, and people will bet faster on roulette.

Some time ago, I found myself in the middle of this muddled-up world, while covering a conference for *New Scientist* magazine. Giddy with jet lag and having spent the entire day in a windowless meeting room, I was desperate to spend the few spare hours I had soaking up some sunlight. It was October – which meant that the sun's ferocious heat was muted somewhat – and the desert sky was cloudless, yet the entire city seemed set up to hide this fact, with chains of underground malls linking one hotel to the next so that you never have to step outside.

Finally, I found myself surrounded by mock-Greco-Roman architecture in the labyrinthine mall of Caesar's Palace, glimpsing what appeared to be daylight up ahead. My excitement was quickly deflated as I drew closer and looked up: above me soared an impressive – yet completely artificial – sky. As I slumped, defeated, next to a replica of Rome's Trevi Fountain, it struck me just how perverted our relationship with natural light has become.

* * *

Our biology is set up to work in partnership with the sun. Life itself arose on earth because its relationship with the sun was a special one. Earth's distance from the sun, not too close and not too far away, meant that the water on its surface remained liquid, whereas on Venus it was baked away, and on Mars it was locked up as ice. Sunlight-driven reactions may also have provided the molecular raw materials necessary for the evolution of life in these early oceans. Some 1.4 billion years later, tiny single-celled organisms called cyanobacteria had evolved, which clumped together, forming brilliant blue-green rafts. And though they were individually tiny, they achieved extraordinary feats: taking the sun's light and turning it into chemical energy through photosynthesis, which they stored away as sugar, thus incorporating sunlight into their very being. In the process, they produced oxygen, which accumulated and transformed earth's atmosphere into the hospitable place we know today.

Life flourished and grew more diverse, evolving and changing until, after another 2.4 billion years, the human species branched off. As we feasted upon the abundant plants and animals, and walked under the sun's rays, we too assimilated starlight into the very fabric of our beings. For every plant we ate depended on the sun's energy to grow, as did every animal: these creatures couldn't survive without eating plants – or eating animals that ate plants.

And as the sun's light penetrated our eyes, it changed the chemistry of our brains, tweaking pathways that control our internal sense of time. So the sun brought order to our ancestors' biochemical reactions and behaviours and, as they looked up at the sun and at the pinpricks of light that decorated the heavens, they found that it also brought spiritual order to their lives.

Little wonder then that humans have long worshipped and revered our nearest star: from the Stone Age solstice worshippers of Britain and Ireland to the Inca who believed they were descended from the sun god Inti. Our histories, religions and mythologies are packed with solar symbolism – whether it's the Greek god Helios pulling the sun across the sky in his chariot; the ochre-hued Sun Woman in the mythology of the Aboriginal people of northern Australia, carrying her torch across the sky; or the significance of light and rebirth in Christianity.

This makes sense, because from humanity's very beginnings, the sun has governed both our bodies and our experience of the world. To our ancestors, the daily rising and setting of the sun, and the seasonal fluctuations in heat, light and food must have seemed extraordinary, not to mention life-changing.

Imagine yourself as a Stone Age man or woman. There are no calendars telling you what time of year it is; no almanacs to explain what's gone before. You don't know that the world is round, tilted and spinning, and that it revolves around the sun, which is just one of billions of vast fiery balls floating in a vacuum called space. And not knowing any of this, neither do you understand that the sun will continue to rise and fall each morning, and the seasons turn, until – in approximately five billion years' time – our sun will burn out completely, but not before it dramatically expands and strips the earth of its water, leaving a dead and barren planet in its wake.

Instead, you look to the heavens and imagine a moving cast of characters, each with a story to tell: a great bear; a chained lady; a hero; a water snake. But, most of all, you revere the biggest and brightest of these celestial bodies: the sun, and its cool, pale companion, the moon. Your senses tell you that when the sun is close and present, plants grow, animals reproduce, and you feel warm and happy. When the sun goes away, everyone and everything suffers.

To you, the sun must seem to have a will of its own; one that could potentially be influenced by your own actions. Because of this, you track its movements, noting where this powerful being rises and falls each day. Its regular disappearance and magical rebirth each morning chime with your own observations of human death and birth; perhaps its circularity gives you hope that we will similarly be reborn one day.

Particularly in a place like northern Europe, you will have watched the sun moving a little further along the horizon each day, as if departing, and associated this with the growing cold, shrinking light, and the withering of your crops to nothing. Finally, for a few short days during the coldest, darkest, deadest time, the sun stops still in its tracks, almost as if it's reconsidering its path (the word 'solstice' translates as 'sun standstill'). Possibly, there's an opportunity to win back its favour. If the sun returns, your seeds will grow; your cattle, pigs and sheep will bear new offspring for you to fatten and consume; and your children will survive. This has happened before, but there's no guarantee that it will happen again.

You don't hold back: people gather from far and wide, animals are slaughtered, and you hold a giant feast; your elders conduct elaborate ceremonies focused around the sun. In the darkness, there springs hope: that the light will return, and life will be reborn from the wastelands.

Archaeological evidence of our ancestors' preoccupation with the solstices – and particularly the winter solstice – has been uncovered at numerous sites, including Newgrange in Ireland, Stonehenge in southern England, Machu Picchu in Peru and Chaco Canyon in New Mexico.

But our ancestors not only revered the sun from a spiritual perspective – they knew that it could be harnessed to promote health. Romans, Greeks, Egyptians and Babylonians all recognised that the sun had powerful curative properties.

Almost 4,000 years ago, the Babylonian king Hammu-
rabi was advising his priests to use sunlight in the treatment
of illnesses. Similar beliefs were held in ancient Egypt and
India, where skin diseases like vitiligo, which causes the
skin's pigment cells to be destroyed, were treated by applying plant extracts and then exposing the affected area to
the sun. Our ancestors clearly noticed that sunlight has the
power to transform seemingly mundane substances, such as
ground-up plant leaves, into agents that can heal.

Such 'photodynamic therapy' has recently been rediscovered, and some skin cancers are now treated by applying
a photosensitising agent to the affected area: when light is
shone on it, a chemical is formed that kills the cancer cells.
Photodynamic therapy is also increasingly used to treat
acne. Meanwhile, modern skin clinics use UV light without
a photosensitising agent to treat conditions such as eczema
and psoriasis because it suppresses inflammation.

Our ancestors used sunlight as a tonic for non-skin-
related conditions as well. The Ebers Papyrus – an Egyptian
medical document dating to around 1550 BC – advised
anointing and exposing painful parts of the body to sunlight. This chimes with very modern research looking at
how the sun's rays affect us: beyond UV light, the sun emits
light from across the spectrum, including the visible wavelengths that become most obvious when sunlight hits a
raindrop, and infrared light. Light from both ends of the
spectrum can influence pain perception: infrared light is
now used as a treatment for various types of acute and
chronic pain and is currently being investigated to promote
wound healing. UV light also stimulates the production of
endorphins, which blunt our perception of pain.

The Greek doctor Hippocrates, who is often referred
to as the father of modern medicine, similarly recommended sunlight for the restoration of health. He promoted

sunbathing and constructed a large solarium at his treatment centre on the Greek island of Cos. Hippocrates believed that sunlight could be beneficial in the treatment of most diseases, although he warned against excessive sun exposure, pleading for moderation – wisdom that still holds true today. In fact, the first accredited description of the deadly skin cancer melanoma came from Hippocrates: its name derives from the Greek words, *melas,* meaning dark, and *oma*, meaning tumour.

Hippocrates also laid the foundations of 'clinical observation', believing that closely watching a patient and recording their symptoms was a critical part of medical care. It was this attention to detail that led him to observe the first recorded example of a daily rhythm in humans besides sleep: a 24-hour ebb and flow in the severity of fever.[1]

Like ancient medics in India and China, Hippocrates also saw the turning of the seasons as important to human health: 'Whoever wishes to pursue the science of medicine in a direct manner must first investigate the seasons of the year and what occurs in them,' he wrote.[2]

Believing that disease arose from an excess or deficiency of four bodily fluids – phlegm, blood, yellow bile and black bile – Hippocrates argued that seasonal changes in these 'humours' explained peaks and troughs in different illnesses at various times of year. He advised people to adapt, according to the seasons, what they ate and drank, the type of exercise they took, and even how often they had sex, in order to keep these humours in balance.[3]

Another celebrated ancient Greek physician, Aretaeus of Cappadocia, recommended that 'lethargics' be exposed to sunlight, while the Roman physician Caelius Aurelianus wrote that light as well as darkness could be used as a medical treatment, depending on the condition. Solaria also featured in many Roman homes and temples, and sunbathing was particularly advised for epilepsy, anaemia, paralysis, asthma, jaundice, malnutrition and obesity.

Although there is no record of any clinical trials demonstrating the effectiveness of such methods, today we can see some plausible mechanisms by which sunlight exposure might have exerted a therapeutic effect. For instance, we know that the sun enables us to manufacture vitamin D in our skin, and that levels of it vary over the year; several studies have linked vitamin D deficiency to epileptic seizures and anaemia. The bone disorder rickets is also caused by a lack of vitamin D, while supplements of it have been shown to help prevent upper respiratory tract infections and the worsening of existing asthma.

Phototherapy is widely used to treat jaundice in newborn babies; light in the blue-green part of the spectrum breaks down the bilirubin pigment in the blood that causes it. Meanwhile, conditions associated with lethargy, such as insomnia and depression, as well as obesity, have been linked to a disrupted body clock, and regular exposure to daylight – particularly first thing in the morning – can strengthen these daily rhythms. Sunlight exposure also boosts the availability of the mood-regulating substance serotonin in the brain, while darkness is being investigated as a treatment for mania.

* * *

Seasonal and daily fluctuations of light and dark – and their impact on our bodies – are increasingly being investigated and accepted by modern scientists as well. We live in a world that is vastly different from the one inhabited by our ancestors, and our lives bring pressures to bear which have a significant impact on our well-being. Humans have evolved to synchronise sleep with when it's dark outside and we're usually at our most active when the sun is up. As anyone who has worked a night shift or flown a long distance and contended with jet lag will know, this isn't a set-up that we can easily override: it's very difficult to sleep when your body thinks it should be awake, and vice versa. But sleep is just the tip of the iceberg. The body is a very different place during the day compared with the night: the kidneys are less active at night, which means that we produce less urine and need to pee less; core body temperature is lower, as are our reaction speeds; and our immune systems respond differently to invaders. Then, as the sun comes up, and day begins, blood pressure and body temperature rise; hunger hormones kick in; and our brains and muscles shift into a higher gear.

These daily fluctuations in our biology are called circadian rhythms – and they are as important to us as they are to the desert coyotes and rattlesnakes that only become active once the sun is low or vanished from the sky: they are the reason we feel so terrible when we're jet-lagged, or start to yawn once the sun has gone down. By tweaking our urges, behaviour and biochemistry, they prepare us for regular events in our environment, like mealtimes or getting up in the morning, which are themselves dictated by the daily cycle of light and dark. Sunlight, and its absence at night, are the main mechanisms we use to synchronise these internal rhythms to the external time of day. If we don't see enough daylight, or we're exposed to too much artificial light at night, our bodies become confused and no longer work as efficiently.

Circadian rhythms begin to develop in the uterus, but those governing sleep don't fully develop until several months after birth. This makes sense: newborn infants need to feed little and often, and a prolonged period of consolidated sleep would interfere with this. Even so, infants receive chemical time cues via their mother's milk, which promote sleepiness during night-time, while infants who are exposed to more bright daylight also sleep more soundly at night.

In adults, there are daily rhythms in body temperature, physical strength, mental alertness, the secretion of various hormones, and many more things besides.

Sunlight doesn't only affect the body clock: it affects our physical and mental health in other ways as well. Most of us are aware that we need regular sun exposure for vitamin D production, which is essential for building a healthy skeleton, but scientists are now discovering new, and astonishing, health benefits to being outside. Mounting evidence suggests that our sun exposure over a lifetime – even before we were born – may shape our risk of developing a range of different illnesses, from depression to diabetes. Recent studies have shown a protective relationship between sun exposure and multiple sclerosis, as well as one with childhood short-sightedness. Being out in the sun, we are starting to understand, can lower our blood pressure, calm our immune system and even alter our mood. Even without such knowledge, most of us are instinctively drawn to sunlight because sitting in it just feels so great, and there may be a reason for that: when the sunlight hits our skin, our bodies release endorphins, the same 'feel good' hormones that produce a runner's high.

There are good reasons why we feel depressed or anxious when we're cut off from the sunlight. As I lurched, like a confused moth, through underground malls and vast casino floors, my sense of time becoming ever more distorted, I considered how we crave the sun in the depths of winter or if we spend too long sitting indoors; how even on gloomy days, taking a walk outside is often a great tonic.

And my mind turned to how a distorted relationship with sunlight can affect, and even harm, our health.

Las Vegas is an extreme example, but most of us have a weaker relationship with the sun than our ancestors ever did. While they were exposed to extremes of light, dark, heat, cold, feast and famine, triggered by the sun, we shield our bodies from sunlight in the daytime, and – thanks to electric light bulbs, screen time and central heating – expose ourselves to an artificial version of it in the evening. This removes many of the natural cues that tell our bodies that it's time to sleep. And because we're more active in the evenings, we're often eating our largest meal of the day when we're physiologically least prepared to deal with it. At the same time, alarm clocks and nine-to-five office hours are waking us up before our bodies are necessarily ready. Besides making us feel groggy and irritable, chronic sleep deprivation is emerging as a major cause of ill health. We need sleep to both mentally and physically recover from daily life: the ubiquity of artificial light at night is robbing us of one of the best preventative medicines going.

Meanwhile, dimly lit offices, sunscreen and indoor living all mean that we're depriving ourselves of the UV light our skin needs to synthesise vitamin D, and – as scientists are increasingly discovering – to tweak our immune systems and help regulate our blood pressure. It also means that we're failing to reap the mood-boosting benefits of sunlight.

But at least nine-to-five office hours are roughly synchronised to the day/night cycle. In 2007, when I travelled to Las Vegas, the International Agency for Research on Cancer added night-shift work to the official list of 'probable' human carcinogens. Being exposed to bright light at night, as both shift workers and casino clients are, forces the body to feel alert when it should be sleeping, setting off a cascade of damaging effects. Shift work – and increasingly bright light at night in and of itself – has been linked to a host of conditions, including heart disease, type 2 diabetes, obesity and depression. Some academics have even suggested that artificial light may be why these conditions have risen to epidemic proportions in modern life. Another theory as to why shift work is associated with so many illnesses is that it encourages us to eat when our bodies think we should be sleeping, which confuses our internal rhythms even further.

Over the past two decades, there has been a scientific revolution in the field of chronobiology – which studies these cyclical changes in our bodies – with the vital importance of our biological relationship with our nearest star becoming ever clearer. In 2017, the Nobel Prize for medicine was awarded to circadian biologists, in recognition of just how important this relationship is to human health. Almost half of our genes are under circadian control, including ones associated with every major illness investigated so far – including cancer, Alzheimer's disease, type 2 diabetes, coronary artery disease, schizophrenia and obesity. Disrupting these rhythms – as we do when we sleep, eat or exercise at the wrong time – is associated with an increased risk of many of these diseases, or a deterioration of symptoms associated with them. What's more, many of the drugs we rely on in modern medicine target biological pathways that are regulated by circadian clocks, which means that they could be more or less effective, depending on when we take them. Meanwhile, side effects associated with radiotherapy and several chemotherapy drugs used to treat cancer can be significantly reduced by delivering them at a time when the healthy cells, which they also damage, are resting.

Our physical relationship with the sun has implications for even the most fit and healthy people as well. World-class athletes are employing circadian biologists to optimise their physical performance, while NASA and the US Navy are applying this astonishing science to keep astronauts and submariners in peak mental shape during their shifts and help them overcome jet lag more quickly.

And it's not only sunlight. Increasingly, we're learning how we can harness the power of artificial light to boost our alertness and physical health rather than undermine it. As we age, our circadian rhythms start to flatten and become less pronounced, so researchers are investigating whether artificial light could be used to supplement daylight in care homes, strengthening these rhythms and alleviating some symptoms of dementia. Hospitals are using circadian-inspired lighting to boost people's recovery from stroke and other serious illnesses, while some schools are using it to boost pupil's sleep, daytime alertness and exam grades.

A better understanding of our relationship with light could improve multiple aspects of our health, both mental and physical: this book will teach you what you need to know, and what you can do to strengthen your own circadian rhythm, optimise your sleep and performance, and hasten your recovery from jet lag. It will also reveal the other health-promoting properties of sunlight, and how to balance this against its damaging effects.

Forging a healthier relationship with light doesn't mean we have to ditch our electronic gadgets and return to the dark ages. But we do need to acknowledge that excessive light at night and an absence of bright light during the day is harmful and so take steps to counter it. We evolved on a rotating planet, when day was day and night was night: it's time to reconnect with those extremes.

* * *

For millennia, then, people have looked at the sun as crucial to our health and seen in its daily and annual cycles a key to understanding the cosmos, yet it is something we seem to ignore or forget about in our modern, daily lives.

Hippocrates would have encouraged us to observe seasonal changes in our mood or energy levels and adjust our behaviour accordingly. However, the relative comfort of our homes and offices, together with the demands of our economic system, encourage us to maintain the same work schedule year-round. We're also expected to maintain a similar level of sociability. Winter is viewed as a gloomy inconvenience and, rather than getting outside to reap what little daylight there is, we switch the lights on and crank up the central heating instead. This may be detrimental to our mental health: exposure to bright light, particularly during the early morning, is a tried and tested way of combating the winter blues. Similarly, many of us keep the lights and heating on long after the sun has set and spend these already brightened evenings in front of electronic gadgets, which produce yet more light. This may undermine our ability to get a decent night's sleep.

The ancients were right to put the sun at the centre of their world. Sunlight was essential for the evolution of life on earth, and it continues to influence our health today. But darkness is also important: the natural cycle of night and day that the sun presides over is implicated in everything from our sleep patterns, to our blood pressure, to our life-spans. Denying access to this cycle, as we do when we cosset ourselves indoors and spend our evenings under bright artificial lights, could have far-reaching consequences that we're only just beginning to grasp.

1

The Body Clocks

THROUGH THE LENS of a solar telescope, the sun appears as a crimson disc against a black background: a banner for a pirate. Keep looking. As your eyes adjust, its surface takes on a mottled, blistered texture. You may notice a dark speck or two, which could easily be dismissed as dust. These are sunspots, darker, colder patches on the surface of the sun, each at least the size of the earth. If you carried on watching for a week or longer, you'd see these specks move across the disc's surface and disappear over its edge. Like earth, the sun is constantly rotating, but whereas it takes our planet 24 hours to complete a full turn, the sun takes twenty-seven days.

The crimson disc itself is as wide as 109 earths, and the photons of light – the very particles that enable your eye to construct this image – have spent some 170,000 years travelling from the centre of this boiling mass of plasma to its outer edge. From there, it took them a mere 8 minutes and 20 seconds of hurtling through space to reach your eye. To lend another perspective, when those photons set out on their journey, man was just inventing the clothing that is shielding your skin from them now.

Mark Galvin has always been fascinated by this aspect of astronomy: that when you observe the heavens, you're gazing back in time. When you look up at the stars on a balmy summer night, you're not seeing them as they are, but as they were hundreds of thousands, if not millions, of years ago. Even the moon is 1.3 seconds older by the time the sunlight reflecting off its surface reaches us, having travelled 250,000 miles through space. Such facts have fuelled Mark's imagination from an early age, and had it not been for his sleep disorder, he would have loved to have studied astrophysics and cosmology at university.

Mark's solar interest is even more fascinating because, unlike most of us, he has lost his biological connection to it. As a result, each day he wakes up about an hour and a half later than he did the day before. After seven days of this, just as Mark's friends and relatives are heading out to work, his body is telling him that it's time to go home. After twelve days, the morning sun is shining through Mark's bedroom window, but his body is convinced that it's midnight. And so it continues, until Mark finally comes full circle and is realigned with normal society. Then the whole cycle begins again.

That's when his body clock is being consistent; sometimes it runs backwards; sometimes Mark will stay awake for 72 hours or sleep solidly for 24 hours. He once slept through an explosion on his street (which saw every house, except his, evacuated on safety grounds) because the police couldn't rouse him.

When I meet Mark for lunch near his home on the outskirts of Liverpool, he is experiencing a rare period of stability. Even so, he texts to warn me that he's running late: his body clock is extremely delayed compared to mine, and he has only just woken up. When he does arrive, he orders a fried breakfast and a cup of tea, while I tuck into a lunchtime sandwich.

Unsurprisingly, Mark's condition – a circadian rhythm disorder called non-24-hour sleep-wake disorder (non-24) – plays havoc with his work and social life. He's been sacked from almost every job he's had because of his terrible punctuality. His friends refer to him as 'Mark who is always late', and he struggles to hold on to romantic relationships; there are only so many times that you can sleep through your girlfriend's birthday, and Valentine's Day, and get away with it. It was Mark who ended his last relationship: 'I was just so fed up of seeing the disappointment on my girlfriend's face, and feeling bad about it,' he says. As for sex, Mark finds that waking someone up at 3 a.m. because he's 'in the mood' doesn't really cut it.

Most people wake up at around the same time each day, even without an alarm clock. If you've ever felt as if you're running on clockwork, it's because to some extent you are. A biological clock is ticking in every cell of your body. All these clocks are driven by the same set of interacting proteins, which are the products of 'clock genes'. You could think of them as the pendulum and gears of a mechanical clock, working together to drive the movement of its hands – only in this case, they drive hundreds of different cellular processes. This enables our cells to keep time with one another, and to synchronise their activities with what they expect to be happening in the world outside.

This analogy could be extended further: just as a grandfather clock with a long pendulum will run slightly slower than one with a shorter pendulum, different people's clocks run at slightly different speeds. Some have short pendulums and run faster: these people tend to be 'larks' who are early to bed and early to rise. Those with long pendulums and slower clocks are usually 'night owls', who like to stay up late and sleep in.

Each of your biological clocks is genetically predetermined to run at the same rate, yet they're susceptible to perturbation by external factors, such as the timing of your meals; when you exercise; the drugs you're taking; or possibly even the activity of the microbes in your gut.

And although there is a clock in every cell, the type of cell influences how responsive it is to these external factors; the clock in a liver cell may be more responsive to the timing of meals, while the clock in a muscle cell may be more responsive to the timing of exercise, and so on.

These clocks do have one very important thing in common, however: all of them respond to signals from a patch of your brain whose job it is to try to keep these clocks synchronised with one other – and with the time of day outside. It's called the suprachiasmatic nucleus (SCN), and it consists of a small cluster of cells in a deeply buried area called the hypothalamus: if you drilled a hole between your eyebrows, you'd eventually hit it. It is closely connected to the pineal gland, which is sometimes referred to as the 'third eye', although this name would really be better applied to this master clock, the SCN.

Containing just 20,000 cells, and no bigger than a grain of rice, this remarkable piece of tissue is the biological equivalent of the Greenwich meridian: it is the reference point used by the billions of other cellular clocks in your body to remain accurate.

If you removed this master clock, as studies in rats and hamsters have revealed, the daily rhythms in your tissues would gradually start to break down. And if you transplanted a new one, those rhythms would reappear – although the innate length of your 'pendulum' would now correspond to your donor's clock. So the SCN really is like a third eye: gazing inwards and outwards, to synchronise internal and external time.

The internal rhythms generated by these cellular clocks are called circadian rhythms – from the Latin *circa*, meaning 'around' and *diēm*, meaning 'day'. They help us to prepare for regular events in our environment, which are linked to the revolutions of our planet. The most obvious example is feeling sleepy when it's night-time, but we're also primed to be stronger, to have faster reactions, and to be better coordinated during the daytime, when we're out exploring the world. Our immune systems may be more responsive to bacteria and viruses then too,[1] and our skin heals faster. There are also daily rhythms in our mood, alertness and memory, and even in our mathematical performance.

Circadian rhythms are thought to have evolved because aligning our activities with the daily light-dark cycle boosts our chances of survival. This has been demonstrated using those same blue-green algae called cyanobacteria, which were so instrumental to the evolution of life on earth (see Introduction). Researchers have engineered mutant strains of cyanobacteria that have far longer or shorter internal clocks than they do in nature. When they kept them in separate flasks, they all grew at the same rate, but when these strains were mixed together, meaning they had to compete for resources, an interesting picture emerged: depending on the length of the light-dark cycle they were grown in, different strains prevailed. When the cyanobacteria were grown under 11 hours of light followed by 11 hours of darkness, replicating a 22-hour 'day', the mutants with shorter clocks outgrew the others; but under a 30-hour 'day', the mutants with long clocks won out. Researchers have also examined how mutants with no circadian rhythm fare: they struggle to compete with the rhythmic strains – except when the lights are permanently switched on.

The clock in cyanobacteria is the earliest ancestral example of a circadian rhythm identified so far. One theory is that such clocks evolved to protect their DNA from sunlight. DNA is extremely susceptible to damage by UV light – just four hours of sunbathing results in approximately ten mutations in the DNA of every skin cell. And although our cells possess enzymes to repair this damage, it's unlikely that they existed billions of years ago in early life forms. DNA is particularly vulnerable when it is being synthesised, so it makes sense to avoid doing this during the daytime, when the sun is at its most intense; indeed, cyanobacteria shut down DNA synthesis for three to six hours during the middle of the day.

Another theory is that cyanobacteria evolved these rhythms to anticipate the daily onset of light-driven photosynthesis, which, although hugely beneficial, also creates reactive oxygen species – also known as 'free radicals' – that are damaging to cells. By anticipating the onset of photosynthesis, cyanobacteria can time the release of substances that soak the reactive oxygen up.

Regardless of why it evolved, the circadian clock fulfils another important function in cyanobacteria today: it separates out competing biochemical processes – some of which are light-dependent – and times them to the most appropriate period of the day or night. Mistiming these events – as would occur if their circadian rhythm was much longer or shorter than the light-dark cycle they lived in – would make them less efficient, which may be why the cyanobacteria with extra-long clocks thrived on long light-dark cycles, whereas those with extra-short clocks thrived on short cycles.

Circadian rhythms are thought to fulfil a similar function in human cells, favouring various biochemical reactions at different times of day; in so doing, they may allow our internal organs to task-switch and recuperate. For instance, a system was recently discovered that drives fluid through the brain while we sleep, flushing out toxins that accumulate during the day, such as beta-amyloid protein, which is associated with the development of Alzheimer's disease. Sleep is also important for the laying down of new memories. These processes don't occur as efficiently when we are awake, so by creating a window when a consolidated period of sleep is actively promoted, our circadian rhythms optimise our ability to learn and recover.

They may also have enabled us to develop as social, group-living creatures. We're more likely to cooperate and work together in cohesive groups if we feel hungry, sociable and sleepy at roughly the same times. Also, as Mark's experience illuminates, we're probably more likely to be reproductively successful, when our libidos are in tune.

Plants have circadian rhythms too: some flowering species open and close their petals at various times of day. During the eighteenth century, the Swedish taxonomist Carolus Linnaeus designed a floral clock based on his observations of when these flowers opened: at 5–6 a.m., morning glories and wild roses opened; at 7–8 a.m., it was the dandelions' turn; from 8–9 a.m., African daisies, and so on.

Gardeners may also have noticed how certain plants are more scented at various times of day: for instance, the Fragrant Cloud rose smells sweetest in the morning; lemon blossom is more scented during the day; stocks and night-blooming jasmine release their heady fragrance in the evening; while petunias – which are pollinated by moths – are more pungent at night. By timing the release of their scent to when their preferred pollinators are most active, the plants save their resources and avoid the inconvenience of the wrong insects drinking their nectar.

It's not only the plants who are in on this game: honeybees are more responsive to visual stimuli during the daytime, when they're out searching for flowers and learn when certain flowers will be open and closed, planning their foraging routes accordingly.[2] Moreover, bees can get jet lag, as demonstrated in 1955, when forty French bees were flown from Paris to New York, where they proceeded to become active and start foraging for nectar even though the flowers they fed on hadn't opened yet.[3]

In fact, circadian rhythms have been found in almost every organism studied so far, from microscopic algae, to subterranean rodents, to kangaroos.

There are a few exceptions: although cyanobacteria and certain other species inhabiting our guts possess circadian rhythms, many bacteria don't; neither do a handful of organisms that have evolved to live in caves, or at the earth's poles. Like those arrhythmic cyanobacteria described on page 20, such organisms probably fare better in these constant conditions because their biology also remains constant. Arctic reindeer are another example: they appear to turn off their circadian clocks during the summer and winter when there is either 24-hour darkness or light. In common with many other animals, reindeer also possess a circannual clock, which means that their biology changes according to the seasons: it is another way of anticipating and preparing for regular changes in their environment. For instance, reindeer and many other species only give birth during the spring, when their young are most likely to survive; reindeer are programmed to grow new antlers then too.

* * *

So how are these daily rhythms generated? The answer lies deep in our DNA, as I discovered when I visited Michael Young's laboratory at the Rockefeller University in New York. There, inside a conical container, brews a microcosm of fruit fly life. At the bottom of the flask, emerging from the pungent, brown sludge, are two tiny, translucent maggots, apparently unperturbed by the swarms of adult flies jostling into one another, launching into flight, and ricocheting off the plastic walls. Also impervious to the chaos are the rice-shaped cocoons that cling stubbornly to the walls of the flask, alongside many motionless adult flies. According to Deniz Top, the research associate who is guiding me through this alien world, these flies are sleeping. You can tell, he says, because their legs are a little more bent, and their heads and bodies are slightly lower than when they're awake. If you touched them with a small stick, you'd have to poke quite hard to get them to move.

Fruit flies are usually creatures of routine: they lay their eggs in the morning, nap in the early afternoon, eat throughout the day, and are most active right before sunrise and sunset. Their larvae also usually hatch at dawn.

But the flies in this flask are 'timeless': because of genetic mutations, they lack a circadian clock. Observing these timeless flies, it strikes me that the chaos in the flask might be what the human world would look like if we didn't have circadian rhythms either. Top looks at his watch: 2.45 p.m. 'At this time of day, most flies would be taking a nap,' he says. He hands me a tube of normal flies, i.e. not genetic mutants: they are indeed largely motionless, and the few that are moving are doing so very slowly.

Young was one of three scientists to be awarded the Nobel Prize in 2017 for piecing together the molecular mechanism of the circadian clock, and they did it by studying mutant fruit flies such as these.

Their work built on studies carried out by Seymour Benzer and his student Ronald Konopka at the California Institute of Technology, during the 1970s.[4] Benzer was fascinated by the strict daily routine of fruit flies and wondered if it might be genetically determined. So he and Konopka began exposing male flies to chemical agents that would mutate the DNA in their sperm, looking for signs of altered timekeeping in their offspring. Eventually, they identified a mutant strain, whose larvae emerged at any time of day or night. Soon afterwards, they identified two additional strains that consistently emerged before or after dawn. All three behaviours were the result of different mutations within what's called the 'period' gene.

Although this suggested that the circadian clock has a genetic basis, it didn't reveal how the clock worked. This baton was picked up by Young, together with Jeffrey Hall and Michael Rosbash, at Brandeis University in Boston, Massachusetts. During the 1980s, they succeeded in identifying several other genes that affected flies' circadian rhythms, including a gene called 'timeless'. They also pieced together how the protein products of these genes drive the circadian clock. Inside every cell, a daily, self-sustaining cycle takes place, which involves these proteins accumulating, coming together, and then switching off their own production, before degrading and allowing the whole process to start over again.

A similar system has since been found to operate in mammalian cells, including our own – and many of the genes involved bear remarkable similarities to those that drive the clock in fruit flies.

The circadian clock is far more than just a biological curiosity; in the two decades since this prize-winning discovery was made, these clocks have been implicated in pretty much every biological process looked at. There is a strong daily rhythm in body temperature; blood pressure; and the hormone cortisol, which (although best known as a stress hormone) promotes alertness – it peaks at wake time and then declines throughout the day. Circadian rhythms govern the release of brain chemicals that regulate mood; the activity of immune cells that fight off disease; and our body's response to food.

Moreover, circadian disruption has been identified as a

feature of every major ailment afflicting society today, from depression to cancer to cardiovascular disease. And it's not only the diseases of rich Western countries that could be better treated if we understood more about these rhythms: the parasite responsible for deadly outbreaks of malaria also times its emergence and development to coincide with its host's body clock, which helps to maximise its spread.

Left to their own devices, these cellular clocks would tick along quite happily at their own genetically determined rhythms. Yet even though some of us have short pendulums and others longer pendulums, most of us manage to survive and thrive on a planet with a 24-hour day. Somehow we remain in check with the daily rotation of the earth, otherwise we'd be like those timeless fruit flies; gradually falling further and further out of synchrony with one another. So, how do we do it?

* * *

During the 1960s, German researchers led by Jurgen Aschoff and Rütger Wever constructed an underground bunker close to the traditional beer-brewing monastery of Andechs in Bavaria, then set about recruiting people to live in it. The idea was to see what happened to people's circadian rhythms when they were isolated from external time cues and free to choose when they ate and slept or switched the lights on and off. The bunker had no windows and was completely soundproofed and shielded against vibrations from traffic. They even wrapped it in copper wire in case electromagnetic forces might influence the ability to keep time.

Inside the bunker were two fully furnished apartments, in which a succession of volunteers lived for several weeks. Food and other items were delivered to and collected from the bunker at irregular intervals, so that the volunteers wouldn't be able to guess the time of day. To keep tabs on the volunteers' rest and activity patterns, the floors of each apartment were fitted with electrical sensors, their body temperature was constantly measured using a rectal probe, and regular urine samples were passed out to the scientists – along with the volunteers' shopping lists. They also kept detailed diaries about how they felt during this experience of living 'timelessly'.

For the first nine days of their stay, light, temperature and noise within the apartment was adjusted to coincide with what was happening in the world outside. Then these external cues were removed, leaving the volunteers to eat, sleep and be active whenever they felt like it.

Isolated from external time, the volunteers continued to spend roughly one third of their time asleep, and two thirds of it awake, but the timing of these daily cycles of sleep and activity varied between individuals: some of them began a new cycle just short of every 24 hours; most of them started afresh closer to every 25 hours. Divorced from the outside world, the volunteers were beginning to 'free-run' on their own internal rhythms.[5] At the time, Aschoff and Wever assumed that it was our social interactions with other people that usually kept us synchronised with the 24-hour world. But it turned out to be something much simpler: light.

It turns out that light acts like the reset button on a stopwatch: it tweaks the precise timing of the master clock (the SCN), ensuring that it remains aligned with the rising and setting of the sun. If you have a long pendulum, then exposure to bright light during the daytime will pull the hands of your clock forward a little bit, so it catches up with the sun; if you have a short clock, it will wind them back a little bit, meaning that everyone stays synchronised.

Light is also what enables us to shift the timing of our clocks when we travel across time zones and the sunrise comes earlier or later. We are particularly susceptible to its effects at night, and shortly after dawn: light in the early evening and at night causes our clocks to delay, so we feel sleepy later, whereas morning light advances the clock and makes us want to sleep earlier the following evening.

It is this very same mechanism that becomes scrambled in people with non-24-hour sleep-wake disorder – the vast majority of whom, unlike Mark, are blind.

* * *

Harry Kennett lost his sight when he was thirteen, after he and a friend discovered an unusual metal object surrounded by small bags of sand in a ploughed field near Minster in Kent. It turned out to be an unexploded anti-aircraft bomb, which blew up when the boys started prodding it. Kennett's friend was killed, and Harry lost his eyes and one of his legs. His injuries might have been even worse if he hadn't gathered up many of the sandbags – ballast for the balloon that transported the bomb – and stuffed them into his dungaree pockets, thinking they might come in useful for his pet budgie. From that point on, in addition to the trauma of the accident, Kennett started experiencing difficulties with his sleep.[6]

Sleep disturbances are common in blind people, but they are both more prevalent and more severe in those who – like Kennett – have no conscious perception of light.[7] Often, such people will experience periods of good sleep, followed by periods of exceptionally bad sleep where they often nap during the day.

The timing of our sleep is regulated by two systems: a 'homeostatic' system which keeps track of how long you've been awake, and gradually builds up a pressure to sleep through the release of sleep-inducing substances in the brain – much like sand accumulating at the bottom of an hourglass – and a circadian system, which sends out alerting signals during the day and creates an optimal window for sleep at night.

Blind people have taught us a lot about how the circadian system works, because they have demonstrated the importance of the eyes in keeping us synchronised with the 24-hour day. If someone loses their eyes, sleep pressure still mounts up, but the sleep window set by their circadian system constantly shifts in line with the length of their internal clock. Some weeks, they'll sleep at night; other weeks, they'll feel sleepy in the middle of the day.

We now know that the reason the eyes are so important to our internal clock is because they contain a very special type of cell, which was only discovered in 2002. Before then, the eye was assumed to contain two types of light-responsive cell: rods, which provide black and white vision in low light conditions and cones, which work in brighter light and enable us to perceive colour.

This assumption was shattered during the nineties, when experiments revealed that mice with a genetic condition which caused their rods and cones to degenerate could still adjust their circadian rhythm system to a changing light-dark cycle, whereas those whose eyes were removed completely could not. Eventually, the mysterious, light-perceiving sentinels were identified. At the back of the eye, nestled behind the layer of rods and cones, lies the master clock's window on the world: a group of light-sensitive cells – called intrinsically photosensitive retinal ganglion cells (ipRGCs) – that enable it to perceive external time. Lose these cells, as you might if your eyes were damaged by a bomb, and your body loses its ability to keep in time with the sun.

When light hits the eye, the ipRGCs send a signal to the brain's master clock, which causes clock gene expression to be altered, and the timing of the master clock to be reset. These retinal cells are particularly responsive to light in the blue part of the spectrum, which includes daylight.

Although it often appears white, sunlight is made up of a broad spectrum of different wavelengths, including blue light. This is not true of many artificial light sources, which tend to be enriched in certain wavelengths and deficient in others. This is important because, although all types of light will shift the timing of the master clock if they are bright enough, some will have a greater effect than others.

For millennia, the only source of light at night was moon- or starlight – which, although it contains a broad spectrum of colours, is very dim – or the light produced from burning wood, wax and oil. Firelight produces a large amount of light in the red part of the spectrum, but very little blue, and it also tends to be relatively dim, which means that its effect on the circadian system is minimal.

Electric light, on the other hand – particularly the LED bulbs found in computer screens, and, increasingly, in ceiling lights and street lights – is far brighter and emits a lot more light in the blue part of the spectrum. This means that far less of it is needed to change the timing of the clock. This is one reason for the recent concern expressed by scientists and medical professionals about exposure to artificial light at night. For more on the topic of how bright light affects us, see chapter 3, 'Shift Work'.

* * *

There is at least one other thing that can tweak the timing of the body's master clock, besides light: melatonin supplements.

Melatonin is a hormone that's released by the pituitary gland at night, in response to a signal by the master clock (the SCN) – for this reason, scientists often use it as a marker of what time the master clock thinks it is. Melatonin is also thought to be one of the key messengers the master clock uses to inform the rest of the body that it's night-time – including those parts of the brain that trigger sleep.

Besides being under the control of the master clock, melatonin release is suppressed by light – particularly blue light. Exposure to artificial light at night therefore shortens the duration of the biological night, which could affect people's sleep, as well as other important processes that occur at night, such as muscle repair and skin regeneration.

As it turns out, the master clock is also itself responsive to melatonin levels. In 1987, the British researcher Jo Arendt published a paper showing that melatonin supplements could be used to shift the timing of the master clock and help people recover from jet lag more quickly.[8] The study created a media furore: 'we had to run to get away from the journalists,' Arendt recalls. However, the upside of all this press attention was that Harry Kennett came to hear about it.

Feeling that his sleep problems had a lot in common with jet lag, he picked up the phone to Arendt and asked if melatonin might help him too.

Intrigued, Arendt agreed to investigate. They arranged that, for one month, he would take either melatonin or a placebo tablet each evening – he wouldn't know which – and then he would be switched to the other treatment for a month. Just two days after starting to take the real melatonin Kennett telephoned Arendt, announcing, 'It's like night and day'. For the first time since the explosion, his sleep was normal again.

Sighted people with non-24, like Mark Galvin, are a different proposition. As a child, Mark slept just fine: it wasn't until he hit puberty that things started to go haywire. At first it was just struggling to fall asleep at night: 'Instead of falling asleep at, say, 10 p.m, it would be 10.15 p.m,' he says. However, very gradually, it crept later. By the age of twelve, he wasn't falling asleep until midnight; and by the time he was fifteen, it was more like 2 a.m. This created problems because he still had to be at school the next morning, so he got progressively less sleep. To make things worse, Mark moved to a new school, which meant getting up at 6.45 a.m. to catch a bus.

He began oversleeping and getting into trouble for being late, and because he was always so tired, his grades started slipping as well. 'I kept being told "you've got such good potential, why don't you try harder?" and "everyone else gets up, why don't you?"' Mark says.

GCSEs were a struggle, but he scraped through and started studying for his A Levels – his sights still set on a career in astrophysics. By now, Mark was struggling to get to sleep before 5 a.m., and was often sleeping for less than two hours per night: 'That's around the time that I started to skip sleep altogether, because I was so worried about not waking up the next day.'

This shift towards later and later sleep during adolescence is common – and is mirrored by a shift back towards earlier sleep in late mid-life to old age – but in Mark's case it was extreme. Puberty seems to trigger a change in teenagers' sleep timing, typically delaying it by around two hours. So asking even a normal teenager to get up at 7 a.m. is a bit like asking a middle-aged adult to get up at 5 a.m.

Exposure to the light from smartphone and computer screens may exacerbate the problem, because exposure to light in the evening delays the clock still further, meaning that teenagers don't feel sleepy until later. However, this nocturnal tendency of teenagers has been seen around the world – even in communities with no access to electricity, where it is weaker, but still present. Not only is their optimal sleep window pushed later, their homeostatic pressure to sleep (that build-up of sleep-inducing substances in the brain) accumulates more slowly, making it easier for them to stay awake.

Adolescence is considered a crucial period for brain development – and teenagers need considerably more sleep than grown-ups: 8.5 to 9.5 hours is considered optimal. Yet a recent US National Sleep Foundation poll found that 59 per cent of young teenagers and 87 per cent of older teenagers were getting significantly less sleep than this – at least on school nights.

The detrimental effects of insufficient sleep have been well documented. Alertness, working memory, organisation, time management and attention span are all impaired; as is abstract reasoning and creativity. Teenagers suffering from chronic sleep deprivation have been shown to achieve lower grades, to have reduced school attendance, and they're more likely to drop out. They're also at higher risk of depression, anxiety and suicidal thoughts; and more likely to engage in risk-taking behaviour such as consuming drugs or alcohol. Studies suggest that teenagers whose parents set – and manage to enforce – earlier bedtimes, are at lower risk of depression and suicidal thinking.

No wonder Mark struggled at school. Eventually he abandoned his university plans altogether and found work in IT – but his tardiness continued, making it difficult to hold down a job. Then, in his early twenties, his sleep changed once again: rather than simply being very delayed, his sleep window started shifting daily.

Mark pulls out his phone and clicks on the app he uses

to record his sleep. He shows me a chart, which maps out the chunks of time that he's slept for over recent weeks and months. Whereas in most people, these chunks would neatly line up because they sleep at more or less the same time each night, in Mark's case, they spread out diagonally across the screen, like a flight of stairs. Each day he goes to bed and wakes up an hour or so later than the one before.

He likens his condition to being in a different orbit to the rest of society, which makes for a lonely existence, apart from the few short days when he realigns each month. When this happens, 'a communication line opens up, when you can do your shopping and speak to family and friends'.

Mark's breakthrough finally came at the age of twenty-eight, when he was working as a help-desk technician at the local hospital. A friend in the chest lab attended a conference and sat in on a presentation about sleep apnoea – a disorder where people briefly stop breathing during sleep. An audience member asked a question about a different condition, in which the timing of someone's sleep shifted later each night, and she learned its name: non-24-hour sleep-wake disorder.

She passed this information on to Mark, who immediately typed the term into Google, and pulled up a list of symptoms: 'Imagine opening a book and reading a description of the main character, in which it describes your hair, what you're wearing, and what you had for breakfast,' he says. 'Everything it said described the experience of my life for the past fifteen-or-so years. It was word-for-word, from the problems getting worse, to the misdiagnoses.'

Armed with this information, Mark visited his GP, who raised an eyebrow, but referred him to a hospital consultant, who then sent him to a sleep clinic. Finally, at the age of thirty, he was diagnosed with the condition. 'To have a neurologist say: "this is real", after twenty years of people saying "you just don't want to get up; you don't want to go to sleep", was a massive relief,' he says. 'It meant I wasn't mad and I wasn't just being lazy.'

No one knows for sure what causes non-24 in sighted people. To some extent, it may be self-imposed: 'They stay up late, see light late, and push the clock later and later, which triggers this free-running behaviour,' says Steven Lockley, a neuroscientist at Brigham and Women's Hospital in Boston, Massachusetts, who is an expert on the disorder. However, it's also likely that they have some sort of particular biological sensitivity to light, or in the ability of their master clock to latch on to it, which has yet to be identified. Several published cases have described the onset of symptoms following a traumatic injury to the head, or aggressive treatment for cancer, which also hints at a physical cause for non-24.

Another theory is that people like Mark have an excessively long pendulum: studies of sighted people with non-24-hour sleep-wake disorder suggests that their internal clock runs on a 24.5- to 25.5-hour cycle, or even longer. Although light can push or pull the master clock to a certain extent, there are limits, which is why those cyanobacteria engineered to have a long internal rhythm were unable to adapt to 22-hour days.

* * *

Individual differences in pendulum length are linked to people's 'chronotype': their innate propensity to sleep at a particular time. Most people will identify themselves as either 'larks' or 'owls', although in reality, chronotypes can't be split into such neat categories because they lie across a spectrum. An extreme lark would typically fall asleep between 9 p.m. and 9.30 p.m. and wake at between 5.30 a.m. and 6 a.m. if they had the choice, whereas an extreme owl might fall asleep between 3 a.m. and 3.30 a.m. and wake at around 11 a.m or 11.30 a.m. The majority of us are in fact 'intermediate types', preferring to go to bed some time between 10 p.m. and midnight and wake up between 6 a.m. and 8 a.m.

Such preferences are thought to develop early in life: a recent study of two- to four-year-olds found that 27 per cent of them were morning types, 54 per cent intermediate types and 19 per cent evening types, which is similar to the ratio among adults.[9] There is also a strong genetic component, so if you don't like going to bed until the early hours of the morning, you shouldn't be too surprised if your children struggle to fall asleep in the evenings too.

A small number of people lie even further to one extreme or the other. Suzanne Milne is one such person: she suffers from a condition called delayed sleep phase disorder (DSPD) and for as far back as she can remember she has struggled to fall asleep much before 4 a.m. This wreaked havoc during her school and early adult life: often, like Mark Galvin, she'd be so anxious about not waking up in time, that she just wouldn't go to sleep at all.

DSPD affects between 0.2 and 10 per cent of the population, depending on the criteria used to diagnose it – and the sleep deprivation it causes can have serious repercussions. For years, Suzanne only slept for about 15 to 20 hours a week. A single mother from the age of sixteen, she couldn't afford to stay in bed in the mornings and catch up on the missed sleep – she had to either get her son, Connor, ready for school, and then go on to college herself, or go to work.

Eventually, this chronic sleep deprivation caught up with her: in 2012, Suzanne suffered from a series of infections, and then began to lose the sensation in her legs. Her doctors suspected a neurological disorder but couldn't pinpoint a cause, until she mentioned her sleep problems. She was eventually referred to a sleep neurologist, who diagnosed her delayed sleep phase disorder almost instantly. As he described it, her body had reached a point where it simply couldn't handle the sleep loss any more.

At the opposite end of the spectrum are people with advanced sleep phase disorder, who seem programmed to wake up at 4 a.m. or 5 a.m., when people like Suzanne are just getting sleepy. Some of the gene variants predisposing us to these sleep patterns have been identified, and they're remarkably similar to those mutations that cause altered rhythmicity in fruit flies. In the case of DSPD, researchers from Young's lab recently discovered that a mutation in a gene called CRY1 – which is involved in resetting the internal clock in response to light in both fruit flies and humans – is common among some people who suffer from the condition, and it delays their nightly sleep by two to 2.5 hours. There are likely to be other mutations too. And, in the case of extreme larks, a mutation in a gene that's closely related to one of those that causes fruit flies to wake early has been implicated.

Annoying as it may be to lie at either extreme – or indeed, to wake up earlier and earlier as you age – such variability in chronotype could be beneficial to wider human society. David Samson is an anthropologist at the University of Toronto who began his career studying sleep in chimps and orangutans before progressing to humans. In 2016, he won a National Geographic grant to study sleep among members of the Hadza tribe, hunter-gatherers living in northern Tanzania.

The Hadza sleep on animal hides or pieces of cloth on the floors of grass huts, with each hut occupied by one to two adults and several children. Approximately thirty adults make up each camp, although several such communities usually live nearby.

While he was there, Samson was struck by the fact that no one was employed to keep watch while the rest of the camp slept, despite numerous dangers stalking the undergrowth. Possibly, he thought, differences in individuals' sleep preferences might make this unnecessary: if there was at least someone awake for most of the night, then they'd be able to raise the alarm. To test this hypothesis, Samson persuaded thirty-three Hadza adults to wear motion-sensing devices on their wrists for twenty days, which could be used to infer when they were sleeping and awake, and would provide some details about the structure of their sleep.

Samson expected to find several hours each night during which everyone would be sleeping simultaneously but, even in these relatively small communities, this was extremely rare. 'It was quite shocking to us,' he says. Because of the range of ages within the camps – the result of extended family members living together – there was a wide spread in the timing of people's sleep, resulting in near-constant vigilance.[10]

Samson believes that this phenomenon could also explain why humans are so long-lived. 'We're calling it the poorly sleeping grandparent hypothesis,' he says. Previously, researchers have suggested that the reason so many people live long past reproductive age is because of the survival advantage to the group conferred by grandparents helping with child-rearing. Now it seems there may be an additional advantage: keeping look-out.

His research has implications for people with sleep disorders. It suggests that the Suzanne Milnes and Mark Galvins of this world – plus the countless older people who find themselves waking up at 4 a.m. or 5 a.m. each morning – are quite normal; in days gone by, they may even have found themselves extremely valuable members of their wider social group.

2

The Body Electric

HANNA AND BEN KING live in a large, modern house, with a fitted kitchen and bathroom, comfortable beds and hot running water. If it wasn't for the horse-drawn buggy in the garage, and their unusual dress sense, you could be forgiven for thinking that you were visiting any American middle-class family home. Unless, that is, you visited after sunset.

As members of the Old Order Amish, Hanna and Ben follow the code of the 'Ordnung': a Pennsylvania-Dutch term for 'order and discipline', which outlines how they should conduct their lives. One of the things it forbids is connection to the electric grid. This isn't because the Amish are opposed to electricity per se: they're allowed to use batteries to power tools in their workshops, or for some practical household items such as Hanna's sewing machine, which she uses to make traditional patchwork quilts and the family's clothes – all in plain fabrics, and without buttons, which are deemed too 'showy'. They even have a solar panel for recharging the batteries for these items, as well as a large, gas-powered fridge-freezer.

Amish households live off-grid because it's an effective way of keeping the modern, 'English', world out. If you're not connected to the grid then there's no TV or internet; no electric gadgets like smartphones, which they worry would change and fragment their community, as well as leading to a less Godly existence.

It also means no electric lights at night. To see, Hanna's

family instead uses a single, large, stand-mounted propane gaslight, which they wheel between their enormous kitchen and living room. The lamp has a patterned glass lampshade, and a toy gibbon hanging off it, and it provides enough light for the family to cook and eat dinner by, as well as enabling them to read or stay up talking once its dark outside. In recent years, the family has also taken to carrying a battery-powered LED lantern with them when they go to the bathroom or enter other unlit parts of the house, including the bedroom – before that, they used a torch or oil lamp. Even so, once the sun goes down, most Amish households are far darker than the average American house.

There are other differences too. The Amish aren't allowed to drive cars, because this might similarly fracture their community, so instead they walk; push themselves along on sturdy, adult-sized scooters; or, for longer journeys, hitch their horses to a buggy and set out on that. Many Amish men work outdoors – around half of men aged thirty-one to fifty are involved in farming – while their wives usually tend large vegetable patches. Also, in summer, the lack of air conditioning forces families outdoors to seek shade when they're at home rather than sweltering indoors. As a result, the average Amish person spends far more time outside than their non-Amish contemporaries. If you want to know what life was like when we had a more direct relationship with the sun, it's a good place to look.

* * *

The early 1800s marked a turning point in our relationship with light. Before then, people experienced night in the old way, when the only source of indoor light besides firelight was the dim, flickering light of tallow candles or whale-oil lamps, which were unaffordable for many people

41

and therefore used sparingly. People came up with innovative solutions to combat darkness and eke out these feeble sources of light: lace-makers would surround a burning candle with globes of water, to 'magnify' the light; miners working in Tyneside would even carry buckets of rotting fish down the mines, as these gave off a faint bioluminescent glow, which helped them to see.[1] Even so, precise work was difficult in these conditions – especially during the winter months – and fire an ongoing hazard, not least in factories, where thousands of lamps were needed to provide enough light to see. Early candles and oil lamps were also smelly and sooty, and lamps needed regular maintenance to keep them working.

The introduction of gaslight was the first major change. The fuel burned by gaslights was a by-product from the production of coke, which was a popular fuel in homes and factories. It was made by heating coal in large ovens, which drew off the gassy fumes.

In 1802, the innovative engineering and manufacturing firm Boulton and Watt installed gas-lighting at their Soho foundry in Birmingham, where they made steam locomotives. These soon spread to other mills and factories, extending the day, and making shift work, which boosted productivity, a realistic proposition. The light produced by gas lamps was brighter than candles or oil, and considerably cheaper.

In 1807, the first gas street lights were erected in London's Pall Mall; by 1820, there were 40,000 of them in the capital alone, not to mention several hundred miles of underground gas pipes, fifty gasometers – large containers used to store gas – and an army of lamplighters employed to tend the street lights, which they lit using an oil lamp attached to a long pole.

As the popularity of gaslights grew, the evenings were

transformed – at least in the cities, which had gas mains. The word 'nightlife' dates from 1852: in the brightened evenings, cafés and theatres flourished, and window-shopping became a popular evening pastime for the rising middle classes. Gaslights also made it safer to wander the streets at night and were credited with reducing crime.

As Robert Louis Stevenson wrote in his 1878 essay 'A Plea for Gas Lamps':

> When gas first spread along a city, mapping it forth about evenfall for the eye of observant birds, a new age had begun for sociality and corporate pleasure-seeking ... Mankind and its supper parties were no longer at the mercy of a few miles of sea-fog; sundown no longer emptied the promenade; and the day was lengthened out to every man's fancy. The city-folk had stars of their own; biddable, domesticated stars.[2]

Some of these gaslights can still be found in isolated pockets of large cities, such as St James's Park in London and Beacon Hill in Boston, Massachusetts. The warm, flickering glow they give off is quite unlike the fierce, blue-white light of modern LED street lights.

Until the invention of paraffin (kerosene) in the 1850s, provincial towns, villages and farms lagged behind in the dark. The demand for paraffin, distilled from petroleum, helped usher in the age of oil. A large paraffin lamp burned as bright as five to fourteen candles, and these soon became a focal point for provincial families during the autumn and winter evenings. No longer did people have to spend their evenings in darkness; these cheaper, brighter lights made it easier to stay up later reading, sewing or socialising. However, this light was feeble compared to what was about to spark up.

The first record of electricity can be traced back to ancient Greece, where, in around 585 BC, the philosopher Thales of Miletus discovered that if he rubbed amber with a piece of fur, the amber would start to attract lightweight objects such as feathers to itself. Primitive batteries, consisting of a clay jar filled with an acidic substance such as vinegar or wine, and a copper cylinder encasing an iron rod, have also been discovered near Baghdad and have been dated to around 200 BC, although their purpose remains a mystery; archaeologists have suggested that they may have been used in electroplating, acupuncture or hooked up to religious icons to deliver a small shock and a flash of light, if touched.

It wasn't until the turn of the nineteenth century that this mysterious force was harnessed to generate light. In 1802, Sir Humphry Davy discovered that passing an electric current through a platinum filament caused it to glow momentarily. Then, in 1809, he demonstrated the first carbon arc lamp, which worked by passing an electric current between two rods of charcoal. Pulled apart, these charcoal rods created an arc of brilliant blue-white light, far brighter than any gaslight. The rods also glowed incandescently, generating an illuminance of their own.

The challenge was to create a more compact and reliable battery, as well as longer-lasting conductor rods, as the charcoal sticks quickly burned away. The breakthrough came in the 1820s, when Davy's assistant, Michael Faraday, discovered that passing an electric current around an iron bar could turn it into a magnet, and that moving a magnet around a coil of wire could create an electric current. The electric generator was born.

Not everyone was a fan of these carbon arc lamps, however. In his 1878 essay, Stevenson continued:

In Paris ... a new sort of urban star now shines out nightly, horrible, unearthly, obnoxious to the human eye; a lamp for a nightmare! Such a light as this should shine only on murders and public crime, or along the corridors of lunatic asylums, a horror to heighten horror. To look at it only once is to fall in love with gas, which gives a warm domestic radiance fit to eat by.[3]

Arc lamps were considered too intense for lighting the home. However, ever since Davy demonstrated the ability of platinum filaments to glow when an electric current passed through them, people had been trying to come up with a way of sustaining this alternative, 'incandescent', source of light. The challenge wasn't only technical: to be practical for general household use, electric lighting needed to be cost effective and easy to use.

In 1878, Thomas Edison picked up the mantle. It was Edison who said that 'genius is 1 per cent inspiration and 99 per cent perspiration'. He also, famously, boasted about needing no more than three hours of sleep per night – although he was often spotted catnapping. As one associate put it: 'His genius for sleep equalled his genius for invention. He could go to sleep anywhere, any time, on anything.'[4]

No wonder, if he was only getting three hours of sleep during the night. Today, the US National Sleep Foundation advises that adults aged eighteen to sixty-four need to get seven to nine hours of sleep per night (seven to eight hours for those over sixty-five), and that individuals who routinely sleep less than six hours (five hours for those over sixty-five) risk compromising their health and well-being.

It seems appropriate, then, that Edison's most famous invention has been so pivotal in undermining our relationship with the natural light-dark cycle, enabling us to work and socialise around the clock. In 1879, Edison successfully tested the first practical incandescent light bulb, and was ultimately responsible for bringing cheap electric lighting to the masses.

Edison didn't achieve this feat alone: his 'invention factory' at Menlo Park, near New York, was crammed with blacksmiths, electricians and mechanics. Also employed on his staff was a mathematician and a glass-blower. Consisting of a coiled filament of carbonised cotton thread surrounded by an evacuated glass bulb, Edison's light bulbs finally allowed households to generate light at the flick of a switch, without the need for an open flame. They were safe enough that a child could be left unsupervised in a lighted room, and they were cheaper than either paraffin lighting or gas.

In the 140 years since Edison's invention, electric lighting has spread far and wide, transforming the way we live our lives. And it continues to grow ever brighter: a recent study of satellite images revealed that the earth's artificially lit outdoor area is currently increasing by more than 2 per cent a year.

Viewed from space, the spidery networks and nebulous clusters of light could be a mirror of the heavens, but from the ground in these brightly lit areas the real stars are vanishing from sight. Today, two thirds of Europeans and 80 per cent of Americans are unable to see the Milky Way from their homes.

'Imagine if we woke one day and were unable to see the green fields and hills of Wales ... the forests of the Amazon, the mountains of Nepal, or the great rivers of the world,' says British professor of cosmology and culture, Nicholas Campion. 'But that is what we have done, and are doing, with the sky, impoverishing our lives in the process.'[5]

Electric lighting has undoubtedly brought many benefits, but it comes at a cost. The loss of our night skies is one of them. The quality of our sleep may be another.

* * *

Donald Pettit sits in the cupola of the International Space Station, his camera lens poised for sundown. As he flies over earth's dark oceans, he records the brilliant flashes of thunderstorms and the undulating beauty of the aurora borealis. But the real light show begins when the continents roll into view. The splats and trails of light glow like a fluorescent Jackson Pollock canvas: orange speckles emanating from sodium vapour lights; blue-green blotches from mercury lights; and whitish-blue webs from newer LEDs.

Pettit has spent more than a year on board the International Space Station, snapping thousands upon thousands of photos of our planet.[6] Such shots are now being stitched together by the Cities at Night project,[7] which aims to document the extent of light pollution and how it's changing due to the growing popularity of LED street lights.

Urban lights scatter photons in unwanted directions, including upwards into space. This scattered light obscures drivers' vision and wreaks havoc on wildlife. Mesmerised by this apparent daylight in the night sky, insects' life cycles are disrupted, birds' migration is thrown off course, and trees cling to their leaves for longer in the autumn – potentially shortening their lives.[8] Even the reproduction of flowering plants is affected by these artificial suns; by disrupting the behaviour of pollinating insects, their daily appointments with flowers that open and close at specific times are missed.[9]

Artificial light also takes a toll on our sleep. A 2016 study found that people living in areas with elevated levels of light pollution tend to go to bed and wake up later than those living in darker areas. They also sleep less, are more tired during the daytime, and are less satisfied with the quality of their slumber.[10]

For centuries, sleep was regarded as a passive and largely dispensable state, and it is an attitude that continues today: 'Don't sleep any more than you have to,' advised Donald Trump in his 2005 book, *Think Like a Billionaire*.[11] He claims to sleep for just three to four hours per night.

Among sleep scientists, though, there's a growing consensus that getting enough sleep is fundamental to our ability to learn, find solutions to problems and regulate our emotions – as well as reading those of other people. Indeed, the way we sleep may underpin our success as a species.[12] Our emotional proficiency is what enables us to cooperate and build flourishing societies, while our creativity, together with our ability to learn and assimilate knowledge, underpins our technological achievements. All these things hinge upon sleep.

Humans sleep in 90-minute cycles, which are further subdivided into periods of non-REM (NREM) and REM sleep. The first half of the night is dominated by NREM sleep (which is itself divided into light NREM and deep NREM sleep), while during the second half of the night REM sleep predominates – although both types of sleep occur during every 90-minute cycle.

The precise purpose of sleep is still a subject of intense study, but one key function of NREM sleep seems to be weeding out unnecessary connections between brain cells, while REM sleep is believed to strengthen those connections.

In his book, *Why We Sleep*, the neuroscientist Matthew Walker likens the interplay between these sleep states to creating a sculpture out of clay: you start out with an unwieldy lump of raw material, equivalent to the mass of old and new memories that the brain has to work with each night. During the first half of the night, NREM sleep excavates and removes large amounts of superfluous material, while short bouts of REM sleep smooth and mould the basic shape. Then, during the second half of the night, REM sleep works on strengthening and defining these basic features, with just a small amount of input from NREM sleep.

It is through this process that our memories are sculpted and archived. Sleep – specifically, the deep NREM sleep that predominates during the first half of the night – helps to consolidate newly acquired memories, so, if you're cramming for an exam, you need this kind of sleep to make those facts stick.

Meanwhile, short and intense bursts of light NREM sleep, called spindles – which are prolific during the second half of the night, where they punctuate long bouts of REM sleep – seem to be involved in transferring recently acquired memories to longer-term storage. This frees up our capacity to learn and manipulate new knowledge the next day. As we age, we experience fewer of these spindles, which could help explain why our memory for new things tends to deteriorate. It's not only facts that are archived during sleep, but physical skills, such as how to juggle more balls or perform a stunt on your bike. Getting enough sleep is therefore extremely important for athletes – a subject that we will return to in chapter 9.

What about REM sleep? This is the sleep state associated with dreaming; animal studies suggest it may also be when we replay memories accrued during the day. One function of REM sleep appears to be fine-tuning our emotions. If we don't get enough of it, we become less adept at reading other people's facial expressions and body language, so our ability to empathise and communicate suffers. We also become less capable of regulating our own emotions. When researchers selectively deprived healthy young adults of REM sleep, but allowed them plenty of NREM sleep, within three days some of them were displaying signs of mental illness – seeing and hearing things that weren't there. They also became paranoid and anxious. This is worrying in the context of owlish teenagers cutting short their sleep because they have to get up early for school (see chapter 10): it is their REM sleep that will suffer the most.

REM sleep is also responsible for cross-referencing newly acquired memories with the back catalogue of older stored memories in your brain. It is during REM sleep that creative insights and abstract connections tend to be made, which is why sleeping on a problem often provides a solution.

We need all these distinct types of sleep if we're to function as intelligent and emotionally competent individuals. And while it's true that some individuals may need less sleep than others, we're kidding ourselves if we think that routinely getting less than six hours of sleep is okay. When we restrict our sleep, it tends to be REM sleep that suffers disproportionately. But fractured sleep – where we sleep lightly and wake frequently – also erodes the NREM sleep that we get more of at the beginning of the night.

* * *

In recognition of the impact that light pollution might be having on our sleep, the American Medical Association recently issued guidance on LED street lights, which are increasingly replacing older mercury- or sodium-based ones. It advised communities against fitting standard blue-white LED street lights – estimated to have a five times greater impact on people's circadian system than older types of street lights – opting for warmer-coloured ones instead; they also suggested that street lights should ideally be dimmable and installed with shields around them to reduce the amount of light reflected upwards into people's bedrooms.

Some city authorities are starting to take notice. New York and Montreal have altered their plans to install standard blue-white street lights, adopting warmer shades instead. In St Paul, Minnesota, tuneable street lights are even being tested that could allow city authorities to adjust their colour or intensity based on the time of day, weather or traffic conditions.

Meanwhile, in towns like Moffat, a solid-looking, former staging post on the road from England to Edinburgh, the street lights have been fitted with shades to direct their glare downwards. Such measures have earned Moffat the title of Europe's first 'dark sky town'.

I visited Moffat to see how these new lights were working out. As I walked through the town, on a frosty October night, the street lights looked like pin pricks of light, rather than glaring beacons, and, off the main streets, it became dark very quickly. On clear nights like these (which are admittedly rare in southern Scotland), the Milky Way casts a magnificent streak across the inky sky.

Such measures are welcome, but they don't deal with the more personal issue of how we choose to light our indoor spaces. Before Edison's invention, the brightest lights in our homes were gaslights, like the ones used by Amish families, and before that, oil lamps and candles. So what impact is our indoor lighting having on our sleep?

* * *

I arrived at Hanna and Ben's home on the Friday before Memorial Day weekend. I was accompanied by Sonia, another 'English' girl (the Amish refer to all outsiders as 'English', although Sonia is American). The daughter of a psychiatry professor who runs medical studies on this community, Sonia had been handed the keys to her dad's enormous pick-up truck in order to drive me there – her first independent road trip, having just completed high school. We picked Hanna up from an indoor farmers' market, where she sells cheese, along the way.

Although they're not allowed to drive, the Amish can accept lifts. Hanna was pleased to see the truck because it would make the weekend more productive. She pulled out a schedule of yard sales she was hoping to attend the following morning. 'Do you want to come with me?'

Going yard-sale shopping, Hanna warned, would mean an early start – at least as we might regard it: 4.45 a.m. is the time she gets up every day, waking without an alarm clock, having gone to bed at around 9 p.m. the night before.

Among the Amish, Hanna is by no means unusual in rising before dawn. On average, Amish people go to bed and wake up approximately two hours earlier than Americans with free access to electricity – meaning that their waking period is more closely aligned with the solar day.

After a quick breakfast of fried egg sandwiches, we're out of the house and pulling into the parking lot of our first yard sale at 5.30 a.m., where several black, horse-drawn buggies are also hitched. Already, a man with a chin-curtain beard, and the distinctive Amish uniform of straw hat, plain shirt and braces, is firing up a barbecue, and the smell of smoke and grilled chicken intermingles with the sweet smell of desserts. Women wearing ankle-skimming dresses, with black aprons and white head-coverings topping their centre-parted, neatly pinned hair, are rummaging through tables of second-hand clothes and bric-a-brac. The Amish typically have large families – Hanna is one of six, but ten is not uncommon – so there are lots of toys, baby clothes, pushchairs and tricycles. Men's second-hand, black, wide-brimmed hats are $5 a pop and there's a lot of Tupperware for sale.

To some extent, these early starts may be cultural: certainly, there are Amish people who would prefer to sleep later. One of them is Katie Beiler, who runs a plastic-food-storage empire. Katie is up at 4.30 a.m. each day because her husband leaves the house at 5 a.m. but, given the choice, she'd remain in bed until 6.30 a.m. 'It's not that I can't get up early, it's just that I love to sleep in,' she says.

Half-past six may not sound like much of a lie-in, but it's all relative. According to recent studies, more than three quarters of Old Order Amish people are early chronotypes or 'larks', compared to just 10 to 15 per cent of the general population.[13]

This pattern of going to bed early and rising shortly before dawn has a long tradition. Buddhist monks are said to have held their hands up to the morning sun, and if they could see their veins, it was time to get up. The same pattern is found among other communities who live without electric light. For instance, a study which examined sleep among members of the Hadza in Tanzania, the San tribe in Namibia, and the Tsimane in Bolivia, found that they too stay up for several hours after sunset, but go to bed relatively early and wake shortly before dawn, sleeping for an average 7.7 hours per night.[14]

Such studies are of interest because they give some clues about how our altered relationship with light might be affecting our sleep. Not only do people living in some pre-industrial societies go to sleep earlier than us, they also seem to sleep better. Between 10 and 30 per cent of people in Western countries experience chronic insomnia, whereas just 1.5 per cent of Hadza and 2.5 per cent of San participants interviewed said that they regularly had problems falling or staying asleep. Neither group has a word for 'insomnia' in their language.

Sonia's father, Teodor Postolache, and his colleagues

have been studying illuminance levels in Old Order Amish households. We measure illuminance in lux, which refers to the amount of light striking a surface. The full moon on a clear night is 0.1 to 0.3 lux, or up to 1 lux in the tropics – about the same as candlelight. In most Amish homes, the average illuminance during the evening is around 10 lux, which is at least three to five times lower than evening light levels in electrified homes.

The researchers have also discovered that daytime light exposure is much higher among the Amish than for most of us in Western countries, where we spend approximately 90 per cent of our time indoors.

This is important because the amplitude of circadian rhythms – the difference between the peaks and troughs of the various rhythms in our bodies – is reduced if we are exposed to more constant light conditions between day and night. Such 'flattening' of the circadian rhythm has been associated with poorer sleep, and is observed in many illnesses, from depression to dementia (see chapter 8, 'Light Cure').

During the summer, Amish people are exposed to an average daytime illuminance of 4,000 lux, whereas the average Brit is exposed to 587 lux. During winter, the Amish experience lower levels of daytime light – around 1,500 lux but for us indoor-dwelling Brits, the average daytime illuminance is just 210 lux: in other words, our waking hours are approximately seven times gloomier than those of the Amish.

Yet it doesn't necessarily feel gloomy to us because the human visual system, remarkable as it is, is a relatively poor judge of illuminance. Your workplace lighting may seem bright enough, but that's because your visual system has adapted to its surroundings, just as it does when you turn off your bedroom light at night and initially can't see anything but then soon make out most objects clearly.

The illuminance in a typical office is between 100 and 300 lux during the daytime, whereas even on the gloomiest, most overcast winter's day it is at least ten times brighter outside. During the summer, when the sun is higher in the sky and there are no clouds, it can reach 100,000 lux.

In the West, we spend our daytimes in the equivalent of twilight, and then keep the lights switched on well after sunset. Some of us even sleep with a night-light on, while city dwellers often have light pollution from street lights to contend with. It's a far cry from the clearly defined daily cycle of light and dark that humans evolved under.

Being exposed to higher levels of light at night does several things: it delays the timing of our body clock and suppresses melatonin, which means that we feel tired later; when our alarm clocks wake us up the next morning, we are still in sleep mode; and overall we get less sleep. It also means that the daily nadir in mood and alertness, which is biologically programmed to occur shortly before dawn when we're asleep, occurs when we're awake instead.

However, the concern around light at night doesn't centre only around the circadian clock and melatonin suppression. Those same light-responsive cells in the eye that synchronise our circadian rhythms also project to areas of the brain that control alertness. Bright light puts the brain into a more active state – it literally wakes us up. One recent study found that exposure to an hour of low intensity blue light boosted people's reaction times (a measure of alertness) by more than if they had consumed the equivalent of two cups of coffee. When caffeine and light were given together, people's reactions were even faster. This could be good news if we're exposed to bright light during the daytime, but at night it could be further undermining our ability to sleep.

This may be one reason why exposure to electronic

screens in the run-up to bed is bad for us. Another study found that, compared with reading a print book, using an e-reader prolonged the amount of time it took participants to fall asleep, reduced the amount of REM sleep they experienced and left them feeling more tired the next morning.

Adjusting the light settings on your phone or tablet – or installing an app that automatically filters out blue light after sunset – can help. Even so, most sleep researchers advocate ditching screens altogether in the 30 minutes before bed – and ideally for several hours beforehand – because even relatively dim light sources held close to the eyes can inhibit melatonin and may therefore affect sleep.

Bright light affects our bodies in other ways too: it increases our heart rate and core body temperature. Usually these things are at their lowest during the night-time, and although the changes brought about by light exposure are relatively small and short-lived, the long-term consequences of repeatedly raising them at night are unknown.

* * *

Ever since the discovery that light – and particularly blue light – can suppress melatonin and alter the timing of our circadian clocks, evidence has been building that exposure to even low levels of light in the evening and during the early part of the night may be affecting the quality of our sleep. Yet light isn't always malign: there is growing evidence to suggest that exposing oneself to bright light during the daytime can help to negate some of the detrimental effects of light at night – as well as improving our mood and alertness more directly.

So what would happen if we followed the Amish's example and reverted to a more traditional relationship with light?

Kenneth Wright, at the University of Boulder in Colorado, has long been fascinated by how our modern light environment might be affecting our internal timing. In 2013, he sent eight people camping in the Rocky Mountains for one week during the summer and measured how this affected their sleep.[15] 'Camping is an obvious way of removing ourselves from this modern lighting environment and just getting access to natural light,' he says.

Before the trip, the average bedtime of the participants was 12.30 a.m., and their wake time was 8 a.m., but both had shifted approximately 1.2 hours earlier by the end of the trip. This was even true of the night owls, who began to look decidedly larkish after a week outdoors. They weren't getting significantly more sleep – at least when the experiment was conducted in summer – but their sleep was more in line with the natural light-dark cycle outdoors. The participants also started releasing melatonin some two hours earlier once they were removed from artificial evening light, and, by the time they woke up, melatonin production had switched off, whereas at home it continued for several hours after waking. Wright suspects that this melatonin hangover could contribute towards feelings of grogginess in the morning.

He recently repeated the experiment in winter.[16] This time, he found that participants went to sleep some 2.5 hours earlier after a week of outdoor living, compared to normal, yet they still woke up at roughly the same time, which meant that they slept for around 2.3 hours longer. 'We think it's because people were going back to their tents earlier to get warm, so they were giving themselves a longer opportunity to sleep,' says Wright, who accompanied them on the winter trip: 'One of the nights it was so cold that we didn't even have a campfire.'

However, the Amish too appear to sleep for around an

hour longer during winter compared to summer. It is still unclear why these seasonal differences in sleep occur – or if it matters if we override them, as we do in modern society.

* * *

Inspired by Wright's studies and observations of more traditional societies, I decided to go cold turkey on artificial light at night myself, and to spend more time outdoors during the daytime. I was interested to see if this would translate into any wider benefits to my health and well-being.

Working with the sleep researchers Derk-Jan Dijk and Nayantara Santhi at the University of Surrey, we designed a protocol to measure the effect of these changes in light exposure on my mood, alertness and sleep. It would be a bit like Wright's camping experiment except that I'd be doing it while trying to juggle an office job and busy family life in central Bristol.

Before the experiment, my sleep routine was fairly typical for a British person: I'd go to bed at around 11.30 p.m. or midnight and be reliably woken up at 7.30 a.m. each morning by my children, who are like human alarm clocks. Even though I slept soundly compared to many of my countrymen – the average British adult goes to bed at 11.15 p.m. but gets just six hours and 35 minutes sleep per night – I often felt groggy in the mornings and would have liked to sleep for longer.

Also, like three quarters of British adults, I had the unfortunate habit of routinely checking my smartphone just before bed, blasting myself with a dose of blue light, which – as we've already learned – inhibits melatonin and pushes the master clock later, potentially making it harder to get to sleep.

Larger studies in the more controlled environment of

a sleep laboratory had hinted that by changing my light exposure patterns I might feel sleepier earlier and fresher in the morning – but this didn't necessarily mean that these benefits would translate into real life: 'We've done a lot of experiments where we've given a dose of light and seen that it shifts the clock,' says chronobiologist Marijke Gordijn at the University of Groningen in The Netherlands. 'If we want to apply those findings to help people, we need to know that it will have the same effect when the environment is more variable.'

Despite the lure of better sleep and happiness, persuading my family to embark on such an experiment took some effort. My husband rolled his eyes, and my six-year-old daughter was only brought round by the promise that it would be just like camping – and by the added bribe of marshmallows.

During the first week, I'd do everything I possibly could to maximise daylight exposure: moving my desk next to a large, south-facing window, loitering in the park after school drop-off, eating lunch outdoors, and substituting indoor exercise with an outdoor equivalent. Another week, we turned the lights off after 6 p.m., even though this meant cooking in the dark – I embarked on the experiment in midwinter. Computers and smartphones were banned in the evenings, unless absolutely necessary, and then only if they were in 'night mode' in order to reduce the amount of blue light they emitted. During a third week, I combined both sets of interventions – keeping things bright in the day and dark at night.

To track my responses, I wore a device on my wrist that captured information about light exposure, activity and sleep. I filled in daily diaries and questionnaires to record my mood and alertness, and I did a battery of online tests to measure my reaction speed, attention and memory. Finally, on the last night of each week, I sat in the dark, spitting into a tube in order to work out when I started releasing melatonin – that marker of internal time. Such is the glamorous life of a scientist.

Cooking by candlelight was a daily challenge. On New Year's Eve, we hosted a candlelit dinner party and managed to undercook our friends' burgers; chopping carrots was an outright hazard. I began preparing meals earlier, which ate into my work time, and panic-checking my pockets to ensure that I hadn't misplaced the box of matches. My pledge to avoid artificial light also made socialising difficult.

Despite the challenges, I did significantly reduce the amount of light I was exposed to after sunset – and this did throw up some interesting findings. During my 'dark weeks', the average illuminance in my home between 6 p.m. and midnight was 0.5 lux – which is only a little brighter than moonlight. Candlelight was perfectly adequate for reading, writing Christmas cards and socialising – and to make dinner preparation a little easier, we eventually installed a dimmable colour-changing light bulb near the cooker.

And, once we adapted, we found that living without artificial light was a pleasure. The candles made the dark winter evenings feel cosier, and conversation seemed to flow more freely. Rather than habitually switching on the television, we turned to more sociable activities, such as board games. Seeing our enthusiasm for this new way of living, friends started dropping by in the evenings to experience it for themselves; they commented on how relaxed they felt in the warm dim light. On New Year's Eve, rather than raucous merry-making, we sat in the dark and played a German board game called Shadows in the Woods (Waldschattenspiel), in which participants take on the role of dwarves who must hide in the shadows of 3D cardboard trees to avoid being caught in the glare of a malevolent tea light. Another bonus was that our children seemed to settle more easily in the evenings (although we didn't quantify this).

Spending more time outdoors in daylight provided another revelation. Initially, it was hard to overcome the belief that, because it was winter, it would be cold and miserable outside, but I was reminded of something a Swedish friend used to say: there's no such thing as bad weather, only inappropriate clothing. And I soon realised that it's rarely as bad outside as it may look. Indeed, the more I did it, the more I came to regard getting outdoors in winter as a treat, rather than a chore.

My attitude to winter began to change. I registered the beauty of hoar frost on rosehips, and the tranquillity of an empty park on a bright December morning, with its long shadows and the sunlight glittering off the ice crystals on the grass.

On one such morning, I took a cup of tea to the park, sat on a chilly bench and made my to-do list for the day. When I pulled out my light meter, it wasn't far off the sort of reading you'd expect on a cloudless day in summer. Back indoors, I took another reading from the middle of my office – it was *600 times dimmer*.

British employers have a duty to provide lighting that's safe and doesn't pose a health risk, but currently this doesn't take the potential impact on our circadian systems into account. The UK's Health and Safety Executive recommends an average illuminance of 200 lux in most offices, while for work requiring limited perception of detail, including most factories, it is just 100 lux.[17] A recent study found that American adults spend more than half of their waking hours in light even dimmer than this, and only around a tenth of their time in the equivalent of outdoor light.

But did doing any of this have any measurable impact on my sleep or mental performance? There was a general trend towards earlier bedtimes. Although, because it was December, social commitments meant that I sometimes ignored my sleepiness and stayed up later anyway: living by the body clock isn't always as straightforward as it is in a lab study. Possibly because of this, the overall amount of sleep I got each night didn't vary significantly between normal and intervention weeks.

Even so, tests showed that – like the participants in Wright's camping study – my body started releasing the darkness hormone, melatonin, some 1.5 to two hours earlier when I cut out artificial light or got more daylight. I also felt more tired in the run-up to bed.

When I correlated my sleep measurements with the amount of light I was exposed to during the daytime, another interesting pattern emerged. On the brightest days, I went to bed earlier. And for every 100 lux increase in my average daylight exposure, I experienced an increase in sleep efficiency of almost 1 per cent and got an extra 10 minutes of sleep.

This pattern has also been seen in larger, better controlled studies than my own. The General Services Administration is the largest landlord in the United States. Its bosses wanted to establish whether designing more daylight into its buildings made any difference to the health of those working inside them. Working with Mariana Figueiro at the Lighting Research Center in Troy, New York, they assessed the sleep and mood of office workers in four of their buildings – three of which had been designed with daylight in mind, one of which hadn't.

The data was initially disheartening. Despite efforts to boost daylight, many GSA workers still weren't getting very much of it: although it was bright close to the windows, once you travelled a metre or so away from them, the daylight largely disappeared. Office partitions and people pulling the blinds further reduced its penetration into the office.

Yet when Figueiro compared workers receiving a large amount of light that was bright or blue enough to activate the circadian system during the daytime – a high circadian stimulus – with those receiving a low stimulus, she found that the former group took less time to fall asleep at night and slept for longer. Exposure to bright, morning light was particularly powerful: those exposed to it between 8 a.m. and noon took an average of 18 minutes to fall asleep at night, compared to 45 minutes in the low light exposure group; they also slept for around 20 minutes longer and experienced fewer sleep disturbances. These associations were stronger during winter, when people may have had less opportunity to receive natural light during their journey to work.[18]

Meanwhile, Gordijn has recently assessed the effect of daylight on sleep structure in a highly controlled laboratory setting and found that it was associated with greater amounts of deep sleep – which you need to feel refreshed in the morning – and less fragmented sleep.[19]

Our sleep isn't the only thing that's affected by daylight exposure. During all three intervention weeks I felt more alert upon waking than normal – but particularly during the two weeks when I was exposed to more daylight.

A recent German study suggested that exposure to bright light in the morning boosted people's reaction speeds and maintained them at a higher level throughout the day – even after the bright light had been switched off. It also prevented their body clocks from shifting later when they were exposed to blue light before bed.

This is good news because it suggests that we may not

need to completely forgo electric lighting in the evenings in order to reap the benefits of improved sleep and daytime performance. Mounting evidence suggests that just by spending more of our daytimes outdoors or exposed to brighter indoor lighting we might achieve the same result. 'When we're talking about the problem of kids looking at iPads in the evening, it's having detrimental effects if they're spending their daytimes in biological darkness,' says Dieter Kunz, who carried out this research.[20] 'But if they're in bright light during the day it may not matter.'

It may even improve their school performance, as teachers at a school in Hamburg, Germany, discovered, when they participated in a study on the impact of different kinds of lighting in the classroom. When the teachers switched on lights that mimicked daylight in both colour and intensity, their pupils made fewer errors in a concentration test, and their reading speed increased by 35 per cent.[21] A study in office workers similarly revealed that exposure to blue-enriched lighting during daytime hours boosted people's subjective alertness, concentration, work performance and mood; they also reported better quality sleep.[22]

* * *

There's another reason why Amish are an interesting population to study from the perspective of light. Lancaster County, where Hanna and Ben King live, is on roughly the same latitude as New York, Madrid and Beijing. Yet, while the prevalence of seasonal affective disorder (SAD) in New York is 4.7 per cent, the Amish have the lowest prevalence yet recorded in *any* Caucasian population.[23]

They also have very low rates of general depression. In part, this could be because of their culture of 'Gelassenheit', or one's 'submission to a higher power'. Acknowledging that you're feeling low could be interpreted as ingratitude for what God has provided, or a preoccupation with one's self.

However, it could also be related to their relationship with light. Because their body clocks are more closely aligned with the solar day, it's likely that biological night has ended for most Amish people by the time they wake up – even though they wake earlier, their master clocks have already issued the orders that cause our mood and alertness to increase during the day. And, because they walk or scoot to work and spend more time outdoors generally, any residual melatonin in their systems that could be making them feel sleepy is suppressed by the bright light.

There could be another reason too. Those light-responsive eye cells that speak to the brain's master clock and the alertness centres of the brain plug into yet other regions that regulate mood. Early morning exposure to bright light is a proven strategy for treating SAD – and there's mounting evidence that it is effective in general depression as well (see chapter 8). Similarly, in the GSA study, those workers who were exposed to a high circadian stimulus during the morning also scored lower on a self-rated scale of depression.

In other words, their early morning starts, combined with walking or scooting to work, and spending a greater proportion of their days outdoors, could be furnishing the Amish with a natural antidepressant.

This also fits with the results of my own experiment. Immediately after waking and before going to bed each night, I filled out a questionnaire to assess how positive and negative I was feeling. This revealed that my early morning mood was significantly more positive during the intervention weeks compared to normal ones. The early morning grogginess was gone: I felt more energetic and uplifted, and ready to start the day. Because of this experience, I'm a convert to outdoor exercise, and I've even come to look forward to certain aspects of winter, particularly the bright, frosty days and the spectacular sunsets.

Taken together, such results emphasise the importance of daylight. They also have significant practical implications. Although few of us would be prepared to spend our evenings in candlelight on a permanent basis, spending time outdoors during the daytime may be something more of us can work into our lives.

3

Shift Work

THOMAS EDISON once commented that, 'Everything which decreases the sum total of man's sleep, increases the sum total of man's capabilities. There is really no reason why men should go to bed at all.'[1]

Although it is true that working 24/7 provides many benefits for society – and the availability of cheap, bright artificial light has made it far easier to achieve this – Edison was wrong about this last point: chronic sleep deprivation is deadly.[2] In some cases, its toxic effects may take years to reveal themselves, but it can also incapacitate us so fast and so seriously that it kills instantly.

An estimated 20 per cent of accidents on British roads are sleep-related, and these are more likely to result in death or serious injury than other types of accident, according to the British road safety charity Brake. Going just 19 hours without sleep – equivalent to waking at 7.30 a.m. and driving home from a party at 2.30 a.m. – puts your attentional abilities on a par with being over the legal drink-driving limit in England and Wales, even if you haven't touched a drop of alcohol.[3] A separate study showed that driving on four to five hours of sleep quadruples your risk of crashing, compared to drivers getting the recommended seven hours sleep.[4]

However, sleep deprivation winds its tentacles around pretty much every physiological process going. It affects our emotional stability, memory and our reaction speed;

and our hand–eye coordination, logical reasoning and vigilance also suffer. Chronic inadequate sleep precedes the onset of Alzheimer's disease, cancer and various psychiatric illnesses; it is also associated with heart disease, obesity and diabetes. It affects the release of both male and female reproductive hormones, and may result in reduced fertility.

In part, these risks relate to missing out on the restorative impact of sleep – simply the number of hours you notch up (or don't) on your bedpost. However, for every one of these diseases and deficits, similar links have been made to disrupted circadian rhythms – and not only because of their effects on sleep.

In one recent study,[5] researchers compared the physical effects of sleeping for five hours per night for eight days in a row with getting the same amount of sleep but at irregular times. In both groups, people's sensitivity to the hormone insulin dropped and systemic inflammation increased, escalating the risk of developing type 2 diabetes and heart disease. However, these effects were even greater in those who were sleeping at irregular times (and whose circadian rhythms were therefore knocked out of alignment): in men, the reduction in insulin sensitivity and increase in inflammation doubled.

Some of the strongest evidence for the harmful effects of circadian disruption has come out of research on shift workers. People who work the night shift are estimated to lose between one and four hours of sleep each day, which is worrying, when you consider the responsibility carried by some shiftworkers, such as doctors, nurses and pilots. But they are also plagued by disturbances to other circadian rhythms.

While shift workers are particularly at risk, few of us manage to maintain our circadian rhythms exactly as they should be. The availability of bright light at night delays our body clocks and alters our alertness, encouraging us to stay up later, even though we must still go to work or school the next morning. As a result, many of us are waking up at a time when our bodies still think they should be sleeping, and then lying in at weekends to catch up on missed sleep, which changes our light exposure yet again. Harmless as this may seem, the 'social jet lag' caused by these inconsistences is akin to physically travelling across several time zones each week. It is also extremely common: studies by Till Roenneberg at the Ludwig Maximilian University in Munich, who came up with the term 'social jet lag', and has interrogated the sleep timings of more than 200,000 individuals from around the world, have concluded that just 13 per cent of us are social-jet-lag-free; 69 per cent of people experience at least one hour of social jet lag per week, while the remainder suffer from two or more hours.[6]

These are more than just numbers: another recent study found that for every hour of social jet lag people experience each week, their chances of suffering from cardiovascular disease increases by 11 per cent and they experience worse mood and greater levels of tiredness. Adding an hour of social jet lag to your week also boosts your odds of being overweight by a third.[7] So, perhaps it's no wonder that the socially jet-lagged among us are also more likely to smoke and drink heavily.

In Roenneberg's words: 'The more of it you experience, the fatter, dumber, grumpier and sicker you will be.'[8]

To understand more, and how we could find some solutions, I talked to someone who has spent almost their entire career secluded from natural light.

* * *

Life on a submarine is stressful. The pressure from the

ocean water outside is crushing. So, in order to minimise the risk of leaks, submariners must pass through several hatches before reaching the living quarters, which eats up living space. Space is further restricted by all the equipment subs carry with them: a nuclear reactor to generate energy; machines to distil drinking water and purify the air; an arsenal of torpedoes; and all the food the crew will need to survive for months beneath the waves. Submariners work in shifts, so someone is always sleeping, which means that the light in the berthing area is constantly dimmed. It's also low in the control room so that the periscope operator can maintain night vision (it has been claimed that pirates, who often attacked at night, would wear an eye-patch to achieve the same result).

It's small, and cramped and dark; and it smells of musty, recycled air and diesel; 'ship smell', as it's known among submariners and their partners. Except for the periscope operator, the hundred-odd men who willingly cram themselves into this unforgiving environment often don't see sunlight for months on end.

If they're in the right part of the world and it's safe, one of the things submariners love most is to 'steel beach' – to pop open the hatch, and allow the crew to go swimming, smoke cigarettes and light a barbecue on the topside of the submarine. From a commanding officer's perspective, providing that opportunity can buy a lot with the crew: 'They are so excited to come up, they're like little kids,' says Captain Seth Burton, a submarine commander in the US Navy. 'But you have to wear your sunglasses, because you have all these submariners that have seen no sunlight – and they're just so pale white when they come out with just their skinny little swimsuits on.'[9]

The usual concepts of day and night are meaningless when you're at sea: there's no sunlight and, because everyone is working shifts, there's no 'normal' society that you're constantly trying to slot back into. But shift work can play havoc with your sleep and health all the same.

When Burton joined the navy, US submarines operated on an 18-hour 'day': submariners would stand watch for six hours; spend a further six hours 'off-watch', which included some training and drills; and then they'd have six hours to sleep. This essentially meant that a new day came, not every 24 hours, but every 18 hours. The body is unable to adapt to such a schedule: it begins to free run on its internal close-to-24-hour rhythm, while mealtimes and sleep opportunities arrive six hours earlier day after day after day. Although there's no sunlight, there's the additional problem of being exposed to bright light in the mess room – often shortly before sleeping – which the SCN (the master clock) latches on to as a substitute for daylight.

Combined with the stress, and the business of living at close quarters with other men for months at a time, the constant jet lag that this schedule induces makes getting a decent night's sleep near impossible. For the first fifteen years of his career, Burton claims that he routinely got by on four hours of sleep per day. He was constantly exhausted: 'The Watch routines didn't encourage the proper amount of sleep, or a consistent pattern of sleep. You were awake when you were meant to be asleep, and asleep when you were meant to be awake.'

Burton's work schedule was extreme, but the circadian desynchrony it created is similar to that experienced by anyone who routinely fluctuates between day and night shifts or does a large amount of international travel for work. Even those with their feet planted firmly on home soil, but who regularly set an alarm clock for work and then sleep in on weekends, are likely to be experiencing some degree of circadian misalignment – a mismatch between the time in their environment, and the body's internal time – with consequences for their health.

Although submariners are highly trained and are taught about the value of good sleep, sleep deprivation is often identified as a contributing factor when collisions and other serious incidents have occurred. 'A very talented person might make bad decisions, just because they are exhausted,' Burton says.

Burton blames the relentless schedule, the lack of adequate sleep and the high-stress environment for the aggressive cancer that he developed in his chest wall when he was twenty-seven years old. This has never been confirmed, but it's plausible: mounting evidence connects circadian misalignment and shift work to cancer.

* * *

According to European and North American surveys, 15 to 30 per cent of the working population is engaged in some form of shift work, with 19 per cent of Europeans working at least two hours between 10 p.m. and 5 a.m. In the UK, 12 per cent of the workforce – around 3.2 million people – regularly work nights – an increase of 260,000 in the past five years.

Although there may be some who enjoy working nights, for many it is a constant struggle. It's not so bad if you consistently work the same shift and can simply pull the blinds and go to sleep as soon as the night shift is over. But many shift workers have kids to get to school in the mornings, or friends or partners they'd like to spend time with during daylight hours. Even if they don't, just a few minutes of morning light – possibly experienced during their journey home – can counteract and postpone the ability of their internal clock to adapt to the night shift.

More than two thirds of people who work night shifts show no circadian adaptation at all, which means that they're active when the body thinks it should be sleeping; seeing bright light when it thinks it should be dark; eating snacks and meals when the digestive system thinks it should be resting, and then attempting to sleep when the internal clock is firing off alertness signals to switch the body into daytime mode.

Irregular or rotating shifts, where people work one or two night shifts per week, are particularly difficult to adapt to. It's not that our body clocks can't adapt – remember that light at night delays the clock and light in the morning advances it – it's just that it takes time. Generally, the master clock in the brain moves about an hour or two per day when adapting to a new light-dark schedule; whether this is the result of switching from day to night shifts or adapting to a new time zone. This means that, depending on the magnitude of the change, it can take several days, or even weeks, to fully adapt. Compounding the problem, the 'peripheral' clocks in our organs and tissues don't adapt at the same rate – and some of them can be further disrupted by, for example, eating when the body isn't expecting it – so not only do they fall out of synchrony with the outside world, they also fall out of synchrony with each other.

Imagine a bakery production line: to get a decent product, the individual jobs need to proceed in a fixed order. If they cease to be coordinated then, rather than a cake, you could end up with a glacé cherry crumble topped with a fried egg.

So it is with the body. Complex processes, such as the metabolism of fats or carbohydrates from the diet, require the coordination of numerous processes occurring in the gut, liver, pancreas, muscle and fatty tissue. Circadian clocks enable these organs and tissues to predict the arrival of food, so that they can process it as efficiently as possible. They also enable the chemical processes occurring inside them to proceed in the appropriate sequence, rather than all at once. If the conversation between them becomes scrambled, they become less efficient, which can lead to, for example, dangerously high amounts of glucose circulating in the blood. If sustained, this can lead to type 2 diabetes, where the pancreas no longer produces enough insulin – the hormone that allows the glucose in our blood to enter our cells and be used as fuel – and glucose levels climb even higher. Over time, the glucose can damage tissues elsewhere, such as blood vessels or nerves in the eyes and feet. In the worst cases, this can result in blindness, or amputations.

In recent decades, epidemiological studies have associated frequent shift work with some alarming health consequences. Shift workers are more likely to be overweight and suffer from type 2 diabetes. They have a higher risk of heart disease, stomach ulcers and depression. Studies of air cabin crew[10] have associated regular long-haul flights with memory problems and, longer term, with significant shrinkage in brain areas associated with thinking and learning. Animal studies have shown that such brain impairments are not simply the result of sleep loss: a disrupted circadian system was causing fewer neurons to be created, and it's this very process of 'neurogenesis' throughout life that is thought to support the formation of new memories.[11]

Another recent study[12] found that a single night shift altered the circadian rhythms in chemicals produced during the breakdown of foods by the digestive system by 12 hours, suggesting that clocks in the gut, liver and pancreas had undergone a dramatic shift in their timing, even though the master clock in the brain had only shifted by about two hours. Two of these 'metabolites' – tryptophan and kynurenine – are commonly associated with chronic kidney disease.

Long-term shift work is also associated with the development of certain cancers – particularly breast cancer. The theoretical basis for this link was first put forward in 1987 by Richard G. Stevens, now at the University of Connecticut. Researchers had long speculated about why breast cancer is less common in low-income countries and grows more prevalent as countries industrialise. At first, Stevens and his fellow epidemiologists assumed that dietary changes were to blame, but when study after large study failed to confirm this, they drew a blank.

Stevens's 'Aha!' moment came when he awoke one night and was struck by the brightness of his apartment. 'I realised I could read a newspaper by the light coming in through the windows,' he says. 'Then I thought: "artificial light: that's a hallmark of industrialisation".'[13]

Various animal studies had suggested that melatonin might have anti-cancer properties. Aside from its links to the circadian system, melatonin also helps to mop up reactive oxygen species, or 'free-radicals', which are generated by normal metabolism and can damage DNA and other cellular components. If melatonin is suppressed because of regular exposure to bright light at night, it seems likely that more cancer-causing mutations will occur.

In fact, Stevens now believes that melatonin's role in maintaining circadian rhythms is of greater relevance to cancer. The secretion of numerous hormones – including oestrogen, which helps to drive the growth of some types of breast cancer – fluctuates over night and day, and if melatonin is suppressed then their levels would be altered, which might enable tumours to grow more quickly. Clinical studies have indeed suggested that peak levels of melatonin are lower in women with metastatic cancer compared to healthy women, and larger tumours have also been associated with lower levels of melatonin. What's more, totally blind women, whose melatonin secretion is unaffected by exposure to light at night, appear to have a lower incidence of breast cancer.

However, there's more to it than just melatonin, as studies in mice that are unable to produce this hormone reveal: they develop more tumours when they are exposed to light-dark cycles that mimic shift work than do normal mice. The circadian clock controls the body's response to DNA damage, and if these surveillance and repair systems are no longer coordinated with the times of day when DNA damage is most likely to occur, this could lead to cancer-causing mutations being missed and left unrepaired.

In the decade after Stevens first proposed a link between breast cancer and shift work, various epidemiological studies were published that seemed to support it. The first was a large study of Norwegian women, who had worked as radio and telegraph operators between the 1920s and 1980, mostly on merchant ships.[14] Initially, the researchers were concerned about the impact of radio frequency radiation on their DNA; instead, they found an association between long-term shift work and breast cancer in later life.

Further support came from the Nurses' Health Studies carried out in America, which are some of the largest investigations into the risk factors for chronic diseases in women ever to have been done. They too found an association between shift work and breast cancer – as well as colorectal and endometrial cancer – even after controlling for things like body weight, alcohol intake and exercise levels. Other studies have linked shift work to an elevated risk of cancers – particularly prostate cancer – in men. And animal studies have indicated that tumours grow faster in mice whose circadian rhythms have been disrupted.

In 2007, the International Agency for Research on Cancer classified shift work that causes circadian disruption as 'probably carcinogenic to humans'. Their intervention came after twenty-four scientists from ten different countries reviewed the available epidemiological evidence plus the results of numerous animal and cellular studies.

Although they cautioned that it was limited, and that more research was needed – particularly to identify the most harmful types of shift work – they found the evidence for a plausible link between disrupted circadian rhythms and cancer 'compelling'.

Two years after the IARC classification, the Danish government began offering compensation to women who had developed breast cancer and had a history of working shifts. Even so, the association between shift work and cancer remains controversial: Seth Burton will never know if those years of working 18-hour days in the dim twilight of a submarine really caused his cancer. Following his diagnosis, Burton underwent surgery and became a health nut, eating 'a lot of wheatgrass', and ditching meat; he also read a lot about circadian rhythms and started prioritising sleep. In June 2018, Burton marked his nineteen-year anniversary of being cancer-free.

Despite his experience, two years after the surgery, Burton went back underwater and is now a submarine commander; as he progressed up the ranks, he became more involved in discussions about the role of sleep and circadian rhythms in submariner performance. In 2013, his submarine, the USS *Scranton*, sea-tested a new 24-hour watch routine during a seven-and-a-half-month deployment, to investigate whether alleviating circadian misalignment might improve sleep and alertness. Among those testing the crew was Mariana Figueiro, director of the Lighting Research Center in Troy, New York: 'Their reaction times were faster and their sleep quality better,' she says.

According to Burton, the crew also started to look physically different: they lost weight and had better muscle tone. He suspects that this is because they were getting more sleep, and because they felt better, they exercised more. However, other research suggests that by imposing a regular routine on their mealtimes, sleep and other daily activities, this could have had other knock-on effects on their health – including weight loss.

* * *

The sleep lab at Brigham and Women's hospital in Boston is widely regarded as one of the top such facilities in the world. One of the first things you notice as you approach it, via a corridor from the main hospital, is that you're walking uphill: the entire floor of the research area is thicker than the other hospital floors, and it floats separate from the rest of the building so that vibrations from everyday life don't reach the research participants and help them decipher what time of day it is. None of the pods where the participants spend their days and nights has external windows, and to enter them you must pass through a double set of doors in order to ensure that no daylight gets in. The technicians who attend them are trained not to say 'good morning', or 'good evening', talk about the weather, or wear sunglasses on their heads – anything that might provide a clue about the time of day is banned. During longer studies (the longest so far is seventy-three days) volunteers can read newspapers, but they are given in a scrambled order, and never on the actual day they're published. Even letters from friends and family are screened, and if necessary redacted, to ensure they give no reference to how much time has passed.

One of the problems with epidemiological studies, such as those investigating the link between shift work and cancer, is that real life gets in the way and it is impossible to control for every factor that might influence the results. But in the highly controlled environment of a sleep lab, many of those factors can be removed. One type of experiment conducted at Brigham and Women's Hospital is the forced desynchrony protocol, which involves exposing volunteers to a 20- or 28-hour 'day' to deliberately decouple internal and external time and investigate the effects of this circadian misalignment on their bodies. Such studies have confirmed that disturbed sleep and decreased vigilance and mental performance are common features of circadian desynchrony – but it's the impact on metabolic and heart function that is currently drawing the most attention.

Frank Scheer didn't set out to become a chronobiologist, but during his undergraduate degree in biology he became fascinated by the human brain; then he encountered the brain's master clock, and its role in regulating the sleep-wake cycle, and he was hooked. Being comprised of such a small number of cells, the SCN seemed to Scheer to be a manageable thing to study. However, the discovery of multiple clocks within the body, each generating its own rhythm, one that can be decoupled by things like food, has made Scheer's research a vastly more complex challenge.

In 2009, Scheer set out to investigate what would happen to a hormone called leptin, which signals to our bodies that we're full after eating, if people's circadian rhythms became misaligned during a forced desynchrony. After just ten days, his ten previously healthy volunteers had deteriorated to the point where three of them met the diagnostic criteria for pre-diabetes. They became less sensitive to insulin, and their blood sugar levels went up; they also secreted less leptin, which left them feeling less sated after eating. What's more, their blood pressure rose by

3 mm Hg, enough to be clinically significant in people with high blood pressure.[15]

His findings could help explain why Captain Burton's men lost weight when they were able to get more sleep, and could eat, sleep and exercise at the same time each day. Sleep deprivation has also been shown to skew the balance of leptin and a second, hunger-boosting hormone called ghrelin, which helps explain why we often want to eat more when we're tired, and tend to crave less healthy sweet, salty and starchy foods.

Mounting evidence suggests that, in terms of broader aspects of health, as well as maintaining a healthy weight, it's not just what we eat, but when we eat it that matters. And that applies to everyone, shift worker or not.

* * *

Gerda Pot is a nutrition researcher, investigating how day-to-day irregularity in people's energy intake affects their long-term health. She was inspired by her grandmother, Hammy Timmerman, who was rigorous about routine. Each day she'd eat breakfast at 7 a.m.; lunch at 12.30 p.m., and dinner at 6 p.m. Even the timing of her snacks was intransigent: coffee at 11.30 a.m.; tea at 3 p.m. When Gerda came to visit, she soon learned that sleeping in was a mistake: 'If I woke up at 10 a.m., she'd still insist I ate breakfast, and then we'd be having coffee and a cookie half an hour later,' she says.

Increasingly, though, Gerda is convinced that Hammy's rigid routine helped keep her in good health until she was almost ninety-five; enabling her to live independently until her final year, and even master Skype so that she could keep in touch with Gerda when she left the Netherlands and moved to London. Using data from a national survey which has tracked the health of more than 5,000 people for over seventy years, Gerda discovered that it's not only what people eat that makes a difference, but consistency in the amount they eat at each meal:[16] even though they consumed fewer calories overall, she found that people who had a more irregular meal routine had a higher risk of developing metabolic syndrome – a cluster of conditions, including high blood pressure, elevated blood sugar levels, excess fat around the waist and abnormal fat and cholesterol levels in their blood, which together increase the risk of cardiovascular disease and type 2 diabetes.

However, when we eat our meals is also important. Scientists have long noticed differences in the way we respond to food at various times of day. When overweight and obese women were put on a weight-loss diet for three months, those who consumed most of their calories at breakfast lost two and a half times more weight than those who had a light breakfast and ate most of their calories at dinner – even though they consumed the same number of calories overall.[17]

Many people think that the reason you gain more weight if you eat late at night is because you have less opportunity to burn off those calories, but this is simplistic. 'People sometimes assume that our bodies shut down when asleep, but that's not true,' says Jonathan Johnston at the University of Surrey, who studies how our body clocks interact with food.[18]

The way we metabolise and process food varies across the day, which makes sense: 'If your food is arriving at a regular time of day, you want your metabolic clocks synchronised to when you're going to eat, so that they can process it as efficiently as possible,' says Johnston.

One thing that varies across the day is the sensitivity of our tissues to the hormone insulin, with people becoming more resistant to its effects at night. Insulin encourages our tissues to take up glucose from the blood, so eating a large meal later in the day could lead to higher levels of circulating glucose. Over time, this might increase someone's risk of developing metabolic syndrome and type 2 diabetes. However, that's not the same thing as gaining more weight. If you're eating more calories than your body uses, your tissues will eventually store some of it as fat, regardless of daily variations in insulin sensitivity.

It also seems to be the case that more energy is used to process a meal when it's eaten in the morning, compared with later in the day, so you burn slightly more calories if you eat earlier. However, it's still unclear how much of a difference this would make to overall body weight. For now, the take-home message is that it is probably healthier to breakfast like a king, lunch like a prince and eat dinner like a pauper – but we don't yet entirely understand why.

So the timing of our clock influences our response to food, but it also works the other way around: Johnston has discovered that the timing of our mealtimes can also alter the timing of our clocks – but not all of our clocks. Altered mealtimes shifted some metabolic rhythms without changing the central clock in the brain.[19] This indicates that mealtime can reset the clocks in human metabolic tissues – perhaps the liver, fat and muscle – which means that eating at irregular times could be another source of circadian misalignment.

Eating irregular amounts at irregular times doesn't just affect our metabolism: the delicately balanced nature of our circadian rhythms means that disruption in one area may have unexpected consequences elsewhere. When mice were fed in the daytime, when they would usually be asleep, scientists found that they sustained more skin damage in response to UV light, compared with those fed at night.

The clocks in their skin had shifted, meaning that a crucial DNA repair enzyme was now being produced at an abnormal time.[20]

There may be other factors, like exercise, that could decouple our clocks if timings are unexpected. Performing vigorous exercise, such as running, just before bed can interfere with your sleep because it boosts levels of adrenaline and cortisol, which increase alertness. However, animal studies suggest that exercising at unexpected times, such as when you'd usually be sleeping or preparing for sleep, also alters the timing of clocks in the muscles, lungs and liver, without altering the master clock in the brain.

The message from these studies is clear: you don't need to be regularly flying across time zones, or working night shifts, for your internal clocks to become scrambled – and, potentially, for your health to suffer. If it's possible to regularise your schedule – which may also mean getting to bed earlier on work nights, cutting down evening light exposure, and trying to get outside more during the daytime – it could have tangible benefits for the way you look and feel. And it may also boost your chances of living to a ripe old age, just like Hammy Timmerman.

Finding a solution to the problem of shift work is less easy. Insisting that people stop working nights is impractical; we need hospitals and power stations to work around the clock, and there are huge benefits to the economy created by both shift work and international travel. Even the sleep lab at Brigham and Women's hospital employs a shift rota to enable its research participants to be monitored 24/7.

However, some of these discoveries about meal timings could help. One thing night-shift workers have control over is when they eat, and if they can keep regular mealtimes during the day, and try to avoid food during the night, they might avoid some of the metabolic disturbances they will be experiencing because of circadian misalignment (at least if they're only working a couple of night shifts a week). This is something Scheer is currently testing.

Another solution to the circadian desynchrony brought about by working nights and absorbing light at the wrong time could be artificial light itself.

* * *

Forsmark Nuclear Power station looms out of the flat forest landscape, like a young child's Duplo model of how a building should look: the three large, grey-white blocks of Forsmark 1, 2 and 3, are each 500 metres high, and topped with straight, 400-metre chimneys. Viewed from the sea, their colour is such that on a cloudy day – as is common in central Sweden – you could easily mistake them for the sky.

Yet there was little mistaking the alarm that rang out at Forsmark on the morning of 28 April 1986. It was triggered by an employee who had accidentally left something in the control room and was going back to retrieve it. En route he passed a radiation detector, which identified high levels of radiation on his shoes, prompting fears that an accident had occurred within the plant itself. Further investigations revealed that he had, in fact, picked up the radiation outside: it had been carried some 1,100 km across the Baltic Sea, from the Ukrainian town of Chernobyl.

Many famous industrial accidents have occurred at night. The Chernobyl disaster happened at 1.30 a.m.; the alarm to the Three Mile Nuclear incident of 1979 was raised at 4 a.m., while the 1989 Exxon Valdez oil spill off the coast of Alaska occurred at midnight. All three incidents involved errors by night-shift workers, which subsequent investigations at least partly attributed to sleepiness.

Our alertness and cognitive performance vary over the

24-hour period, reaching their nadir during the early hours of the morning – at around the same time as our body temperature is at its lowest. They also start to deteriorate if we stay awake for too long, which is bad news, considering that it's not uncommon for people working irregular shifts to go more than 20 hours without sleep, particularly during their first night shift.

The longer the night shift, and the more night shifts worked in a row, the greater the risk. Taking a nap either before or during a shift can help – although, because it can take a while to regain full alertness after sleep, this is a bad idea for jobs that demand an instant reaction to a problem. This rules out mid-shift napping for submariners, who must spring to action at a second's notice. It is also probably unwise if you're manning the control room at Forsmark.

Operating a nuclear power station is a monotonous job, although the managers at Forsmark try to mitigate the potential tedium through education (there are a lot of procedures to learn) and by changing workers' roles. Each day, there is a long list of checks and tests to work through: there are 3,000 rooms in Forsmark 3 alone, some of which you can only enter if you're wearing a radiation-proof suit, or only view through a CCTV camera. And once you get to the end of that list, it's time to start again.

If a problem is detected, you need to be able to think on your feet. The control room operators have instructions on what to do in the event of an earthquake, a flood or an aircraft crash, but they can't plan for every possible scenario. The accident at the Fukushima Daiichi plant in Japan – the result of a 15-metre tsunami disabling the power supply and therefore the cooling of the three nuclear reactors – is testament to that. As Jan Hallkvist, the operations manager at Forsmark 3, says: 'People need to be alert, and able to solve complex problems quickly.'

The control room operators at Forsmark work in rotating shifts, including two night shifts per week. Maintaining alertness within the control room is made even more difficult by the fact that it's buried deep in the heart of the power station, with metres of metal and concrete separating it from the outside world. The problem is particularly severe during the winter months. Located at roughly the same latitude as the Shetland Islands and Anchorage, Forsmark's control room operators see barely any daylight from November to February, regardless of which shift they're working.

As if to compensate for the lack of windows, four paintings depicting the changing seasons have been hung above the entrance to the meeting rooms. But otherwise the control room is a drab, beige kind of place, lined with giant circuit boards that map out the reactors' connections to the grid and show how much power is being generated at any given time.

I'd liken it to a cave, but Hallkvist does it for me: 'We had to do something about the lighting,' he says, walking over to a control panel on the wall.

Hallkvist originally approached the circadian researcher Arne Lowden about his workers' shifts. He was looking for ways to help his staff adjust to their changing schedules and maintain their alertness, but he also mentioned the gloomy control room. Lowden told him: 'If you're going to change the lighting, you should think about circadian lighting.'

Although the high blue-light content of standard LEDs disrupts circadian rhythms when people are exposed to them at night, LEDs also enable at least some of the effects of daylight to be realistically recreated indoors. Because they are tiny, many different-coloured LEDs can be joined together to vary the shade of the light they produce, enabling the colour and intensity of a lighting system to be adjusted according to the time of day.

For an extra couple of thousand euros, Lowden explained, it would be possible to install a 'circadian lighting system' that could supply a shot of intense white-blue light to boost workers' alertness at key times, such as the start of a night shift, but also fade to a dimmer, warmer white in the run-up to the shift's end, preparing them for sleep – in this way, the night shift would be more like an afternoon/evening shift, and when the workers got home they would be ready to sleep. Similarly, the intense, blue-white lighting could provide a substitute for sunlight for those working day shifts in the cave-like interior of the control room, keeping them rooted in the 24-hour world.

Hallkvist was intrigued enough to allow Lowden to test whether such lighting could really boost alertness and sleep in a subset of his workers, as well as help them better adapt to rotating shift work. Beforehand, the illuminance in the control room was a weak yellow light of 200 lux – like that of many offices. The new lights were hung above the reactor operators' desk and, at their peak, yielded 745 lux of intense blueish-white. The rest of the operators worked at desks turned away from the new lights so that they could function as controls.

The experiment was performed during winter, and the results[21] were positive enough to persuade Hallkvist to install the lighting system throughout the control room. The most convincing result was a reduction in the reactor operators' sleepiness during both night and day shifts – but particularly during the second night shift, which is often the hardest. This was despite the reactor operators being exposed to the intense white-blue light for only one to two hours at the start of the night shifts. During day shifts the bright lights were on between 8 a.m. and 4 p.m. – replicating what was occurring in the outside world.

Even so, not everyone is convinced about the wisdom of

exposing night-shift workers to intense white-blue light. It boosts their alertness, but it's also suppressing their release of melatonin and delaying the timing of their clocks. 'It's not an easy fix,' says Scheer. 'There's a risk of making things worse by interfering with light exposure during or after the night shift.' He points to the example of blue-blocking glasses, which some tout as a means of shielding oneself from daylight on the journey home: it's true that this may make it easier to sleep, but if you're driving, it also increases the risk of accidents.

4

Doctor Sunshine

ION MEYER GENTLY LIFTS away the sheaf of white tissue concealing the woman's face. Immediately, it's apparent that there is something terribly wrong with it. The skin is scarred and uneven-looking, and the area around her closed left eye is red and swollen. As I bend forward to take a closer look, I see that the flesh from the bridge of her nose, down to her left nostril, and extending across her left eye socket has been eaten away; through her closed eyelid, the white curve of an eyeball peeks through. A brass plate puts a name to the face:

> Maren Lauridsen
> Lupus vulgaris
> 2.7.18.

This isn't a hospital morgue, but a storeroom at the back of Copenhagen's Medical Museion, and the date refers not to 2018, but to a hundred years earlier.

Today, few people are familiar with Lupus vulgaris, or skin tuberculosis, but a hundred years ago, when Maren walked Copenhagen's streets, it was a disease that was particularly feared. Caused by the same bacterium responsible for tuberculosis of the lungs, it often begins at the centre of the face with painless brown nodules that spread outwards, developing into ulcers that wolfishly consume the flesh (*lupus* is Latin for 'wolf').

There was no cure, so physicians would resort to slowing its spread by burning away the infected flesh with heated irons or corrosive chemicals like arsenic. No wonder people lived in terror of catching it. Once afflicted, victims would become isolated from their friends, family and community, left to face this torture alone.

Although Maren Lauridsen is long dead, this imprint of her disfigured face lives on. Meyer, who oversees the museum's collections, pulls down another crate, and then another; each containing yet more examples of horribly mutilated flesh, immortalised in wax. One face looks as though it was immersed in seawater for several days – it's impossible to tell if it's a man or a woman. Other crates contain just sections of faces: a mouth and jaw reduced to an angry red mush; a pitted, blistered nose.

The models were produced by taking a plaster imprint of the victim's face, then pouring wax molasses into the mould and painting the finished cast. Their function was to document the extent of people's injuries before they underwent a revolutionary new treatment that they hoped would cure them. That treatment was light. Filtered and concentrated through a series of glass lenses and cooled by passage through a water-filled tube, ultraviolet rays were directed onto the faces of the patients, where they set to work killing the disfiguring flesh-eating bacteria.

The man who came up with this treatment was Niels Ryberg Finsen, and he would go on to win a Nobel Prize for his efforts. He would also usher in a new era of interest in the health benefits of sunlight, which continues to this day. Finsen's work had nothing to do with circadian rhythms; it dealt with the direct impact of the sun's rays on bacteria and our skin.

Finsen was born on 15 December 1860, in the Faroe Islands, a small jigsaw of dramatic and improbable-looking peaks that jut out of the North Atlantic some 177 miles north-west of the Shetland Islands. Buffeted by climatic depressions that bring frequent cloud, rain and storms, sunny days would have been in short supply during Finsen's childhood. Perhaps this is what motivated him to try to capture those sunrays and concentrate them, in the process rendering them powerful enough to heal.

Arriving in Copenhagen to study medicine when he was twenty-two, Finsen lived cooped up with his books in a north-facing room where the sun never penetrated. He suffered from anaemia and tiredness – but he noticed that his health improved if he was exposed to sunlight.

In fact, Finsen was in the initial stages of Pick's disease, a progressive disorder characterised by the abnormal metabolism of fats, which begin to accumulate in internal organs, including the liver, heart and spleen, eventually compromising their ability to function. As Finsen progressed through medical school, his conviction about the health-promoting effects of sunlight continued to grow. He collected descriptions about the sun-seeking behaviour of plants and animals and noted how a cat lying out in the sun would repeatedly change position in order to avoid being caught by the shade.[1]

Finsen was particularly inspired by a paper he discovered in an 1877 volume of the *Proceedings of the Royal Society of London*. Written by two British scientists, Arthur Downes and Thomas Blunt, it described an experiment in which test tubes filled with sugar water were left on a south-east facing windowsill. Half of the tubes were left in sunlight; the other half were covered by a thin sheet of lead. After a month, the researchers noted that those tubes that had been exposed to the sun remained clear, while those that had been covered were foul and cloudy. It was some of the first proof that sunlight could kill bacteria;

soon afterwards, the famed bacteriologist Robert Koch – who had only recently identified the bacterium responsible for causing tuberculosis – showed that that, too, could be killed by sunlight.

But these scientists were by no means the first people to take an interest in the sun's healing power at that time. In 1860, the year that Finsen was born, the English nurse and social reformer Florence Nightingale published her *Notes on Nursing*, which contained a section about light. She wrote: 'It is the unqualified result of all my experience with the sick, that second only to their need of fresh air is their need of light; that, after a close room, what hurts them most is a dark room. And that it is not only light but direct sun-light they want.'[2]

Nightingale observed how, in hospital rooms that contained windows, almost all patients lay with their faces turned towards the light, 'exactly as plants always make their way towards the light' – even when lying on that side of their body was uncomfortable or painful to them.

She emphasised how the morning and midday sun (the very times when hospital patients were likely to be in bed) were of most importance. 'Perhaps you can take them out of bed in the afternoon and set them by the window, where they can see the sun,' she suggested. 'But the best rule is, if possible, to give them direct sunlight from the moment he rises till the moment he sets.'

Although the ancient Babylonians, Greeks and Romans had embraced the healing properties of sunlight, for centuries the idea had lain forgotten. Now, in the sun-starved cities of northern Europe, sunlight was being rediscovered. In an era before antibiotics, the revelation that light could kill bacteria constituted a major medical breakthrough. And it was Finsen who developed the first practical application of this discovery.

After graduating from medical school, Finsen found work teaching anatomy in the building that now houses Copenhagen's Medical Museion. His fascination with sunlight continued, nevertheless, and he began experimenting with devices that might harness it more effectively. Today, the shelves of the museum's storeroom heave with the glass and quartz lenses that Finsen developed to conduct his early research into the healing effects of light. He even turned guinea pig himself, quantifying the amount of sun exposure needed to trigger sunburn.

Because sunshine was so often in short supply in Denmark, Finsen also began collaborating with Copenhagen Electric Light Works to develop an artificial light that could be used when the sun was absent. While working there, he encountered an engineer called Niels Mogensen, whose face was blighted with severe and painful ulcers caused by tuberculosis. After just four days of Finsen's light treatment, he showed a dramatic improvement.

From this collaboration, the Finsen Light was born: an elaborate rigging of telescope-like tubes and lenses that filtered, compressed and cooled the rays from a carbon arc lamp, and could be used to treat multiple patients at once. In 1896, Finsen established the Medical Light Institute, which enabled even more patients to be treated, with impressive results: of the 804 people who underwent light treatment for skin tuberculosis between 1896 and 1901, 83 per cent were cured, and only 6 per cent showed no improvement.

Through his experiments, Finsen concluded that it was 'chemical light', identified as blue, violet and ultraviolet light, that was responsible for its healing effects. Initially he thought it was because the rays themselves killed the bacterium that causes tuberculosis, but more recent experiments have revealed that the Finsen Lamp concentrated

UVB rays, which reacted with substances inside the bacteria called porphyrins, causing unstable molecules called reactive oxygen species to be produced that then killed the bacteria.[3] Later, Finsen hypothesised that the light was somehow stimulating the body to heal itself, which may also be true.

By the time Finsen received the Nobel Prize in 1903, his own health had deteriorated to the point where he was in a wheelchair, and he died just a year later, aged forty-four.

Despite his use of electric lights, his passion for sunshine was unwavering, and he frequently encouraged patients to walk about naked in the sun. In an interview conducted shortly before his death he remarked: 'All that I have accomplished in my experiments with light, and all that I have learned about its therapeutic value, has come because I needed the light so much myself. I longed for it so.'[4]

* * *

The nineteenth century was a time of immense change. Not only did it see the invention of new forms of artificial lighting, but the Industrial Revolution also brought flocks of people to the cities seeking work in the mills and factories that were then springing up. A similar thing is happening in developing countries today, and vitamin D deficiency – caused by smog, sun avoidance and clothing which completely covers the skin – is a growing problem, even in sunny countries in the Middle East, Africa and parts of Asia.

Vitamin D is essential for regulating the amount of calcium and phosphorus in our bones, teeth and muscles, and is needed to keep them strong and healthy. Although we get some vitamin D from our diet, mainly from sources such as oily fish, eggs and cheese, most of our requirements are met by manufacturing it in the skin. A substance called 7-dehydrocholesterol absorbs UVB rays from sunlight, converting it into vitamin D_3. This circulates in the blood and is further metabolised into the active form of vitamin D elsewhere in the body. In growing children, vitamin D deficiency causes rickets – characterised by soft, weak bones, stunted growth, and skeletal deformities; in adults it causes the bones to soften, resulting in bone pain, fractures and muscle weakness.

By the middle of the nineteenth century, rickets was widespread in urban Britain and in other rapidly industrialising countries. A survey by the British Medical Association in the 1880s highlighted the urban nature of the problem: rickets was virtually absent from small settlements and agricultural areas. Many who flocked to the booming cities found themselves living in cramped and gloomy conditions, and the burning of coal for the new industries – not to mention the production of gas for lighting – cast a thick blanket of smog that smothered out sunlight and made spending time outdoors unpleasant. Children played in narrow alleyways, between tall buildings, further isolating them from any sunlight that did penetrate through. Add nutritional deficiencies because of poverty into the mix and a legion of bowed, deformed skeletons was the result.

Various theories were put forward to explain the cause of rickets. Jon Snow, best known for his detective work in tracing the origin of cholera to a water pump in London's Soho district, believed that bread adulterated with aluminium sulphate was to blame; this might interfere with the absorption of dietary phosphorus, needed to mineralise and strengthen the skeleton, he said. Others pointed the finger at air pollution.

It was in the late 1880s that an English missionary called

Theobald Palm suggested that sunlight deficiency was the cause. He had recently returned to Cumberland in northern England to practice medicine after having spent ten years in Japan, and he was struck by the contrast: suddenly he was encountering deformed children, something he'd never seen during his time overseas.

After consulting other medical missionaries in China, Ceylon, India, Mongolia and Morocco, Palm became convinced that rickets was a disease of grey skies and gloomy alleyways. He suggested that 'the systematic use of sun-baths' might be a solution.[5]

Combined with the observations of Downes and Blunt about the bactericidal effects of sunlight, and Finsen's success in treating skin tuberculosis with light, Palm's ideas heralded a new appreciation for the sun. It started with wound infections, tuberculosis and rickets, but over the next forty years, 'sun cures' became a mainstay of medical treatment. It was as though sunlight – or more specifically, the UV rays contained within it – somehow boosted health more generally. Exposure to it also made people feel good, and increasingly, society decided, made them look better as well.

* * *

In 1903, the year that Finsen was awarded the Nobel Prize, a Swiss doctor called Auguste Rollier was turning his back on a conventional medical career, following the suicide of a close friend who had become disabled by skeletal tuberculosis. Besides infecting the skin and lungs, tuberculosis can also infect the bones and joints, causing the spine to deform and protrude outwards, or the hip joints to degenerate, resulting in lameness. It was the latter problem that had afflicted Rollier's friend. Parts of his knee and hip joint had been surgically removed while the boys were still at school, but this had failed to contain the disease. Subsequent surgery as a young adult had left him mutilated, and he eventually took his own life.

Soon afterwards, Rollier's fiancée contracted tuberculosis of the lungs. Perhaps in desperation, he turned to a folk remedy that he'd learned from some of his patients: heading to the high mountains and lying out in the sun. In 1903, he accepted a job at a rural surgery in Leysin, in the Swiss Alps, and the couple moved to this sunny village with spectacular views of the fang-like peaks of the Dents Du Midi. It was here that Rollier began to develop an alternative treatment for tuberculosis.

'At an altitude of 5,000 feet,' he wrote in his 1927 book, *Heliotherapy*, 'the air is never oppressively hot, even in the height of summer; in winter, although the atmosphere is intensely cold, the brilliance of the sun more than counteracts this quality.'

On sheltered, outdoor terraces, 'debilitated and miserable patients' would be laid out in loincloths in the sun, 'under conditions that impart to their bodies a more effective means of self-defence than could be possible in the flat country. The sick people regain their lost vital energy through the instrumentality of the sun and alpine air.'[6]

This wasn't sunbathing as we know it today, where the sun-starved masses of northern climates fling their bodies onto Mediterranean sand for a week of intense baking. Instead, Rollier advocated slow and progressive exposure to sunlight, starting with just five minutes on the feet, which would gradually increase over the next three weeks, until all but the fairest patients were taking daily two- to three-hour 'sunbaths' during the summer, and three- to four-hour baths in winter. Believing that the combination of hot air and sunshine was bad for people's health, he banned patients from sunbathing at midday during summer, and preferred to make use of the early morning sun.

Not only did Rollier's fiancée recover, soon many other patients were regaining their health under his supervision. Before and after photos documented the startling transformation of hunched and deformed children's spines reverting to their normal curvature within eighteen months of sun treatment. Others captured men lying sprawled out in loincloths in front of enormous sunny windows, and young boys waving from their beds on sunny outdoor terraces.

In such 'internal' cases of tuberculosis, it seems unlikely that UV light was directly killing the disease-causing bacterium, as it did in the skin. Neither could this bactericidal effect of sunlight explain its role in preventing rickets.

The breakthrough came in 1925, when an American doctor called Alfred Hess discovered that feeding rickety rats the skin of humans or calves that had been irradiated with UV light cured them of their rickets.[7] The mystery curative factor it contained was eventually characterised as vitamin D.

We now know that the reason Rollier's sunlight treatment may have been so effective against internal forms of tuberculosis is because the vitamin D it manufactures helps to turn on the body's first line of defence against bacterial invaders inside the body. When immune cells such as macrophages – which detect, engulf and destroy foreign bodies, including bacteria – encounter an invader, they start converting the inactive precursor of vitamin D into its active form, and producing receptors that will enable them to respond to it. As a result, they spew out an antimicrobial peptide called cathelicidin, which helps to kill the bugs. The same process is thought to reduce our susceptibility to other chest infections besides TB.[8]

By the end of the 1920s and into the 1930s, sunlight was being touted as a cure for, well, pretty much every disease under the sun. In his 1929 book, *The Sunlight Cure*, the medical writer Victor Dane concluded: 'If you wish to obtain a general idea of the powers of the sun, and to know the names of the various diseases it benefits, buy a medical dictionary and memorize the names of all the diseases you find therein. The sun is the great healer of healers, truly an "elixir vitae".'[9] Sunlight had hit the mainstream, and suntans were to become a must-have fashion accessory.

However, not everyone bought into the idea of sunlight as a cure-all. A 1923 paper in *The Lancet* noted that 'the results on tuberculosis of the lungs have been, in many hands, disappointing, and have led to avoidance of the treatment by many physicians, and even to its condemnation by some as a dangerous and unjustifiable form of therapy'.[10] In some cases, unsupervised sunbaths had resulted in an increased temperature, worsening cough, and the coughing up of blood.

Some went even further in their criticism. In his 1947 book, *Nothing New Under the Sun*, the distinguished British surgeon John Lockhart-Mummery dismissed sunlight therapy as 'pseudo-magic', adding that 'most of the benefit patients get from such treatment is due to their faith in the magical results, rather than direct benefit.'[11]

By then, however, the popularity of sunlight as a panacea was already beginning to fade – even if the fashion for bronzed skin that it helped kick-start continued for decades. The discovery of antibiotics rendered sunlight therapy for infectious diseases obsolete, and as smog from the cities cleared, and cod liver oil was identified as a rich dietary source of vitamin D and routinely spooned down children's throats, the threat from rickets also diminished.

Today, phototherapy is still used, but it is confined to certain skin diseases, including psoriasis, atopic eczema, and other forms of dermatitis.

Even so, as concerns about antibiotic resistance grow in our own age, there's renewed interest in harnessing the bactericidal effects of light. Fixtures that use a narrow spectrum of visible indigo-coloured light to kill bacteria are being deployed in hospitals to disinfect surfaces and clean the air. So too is the subtype of UV light called UVC, which is unable to penetrate human skin or the outer layer of the eye yet is deadly to smaller bacterial cells. A recent trial published in *The Lancet* found that UVC machines cut transmission of four drug-resistant superbugs – MRSA, vancomycin-resistant enterococci, Clostridium difficile, and Acinetobacter – by 30 per cent.[12] Unlike antibiotics, which target specific cellular systems, light destroys the nucleic acids that make up DNA, leaving bacteria unable to replicate or perform vital cellular functions.

It isn't only these deadly aspects of UV light that are prompting a renewed interest in sunlight. In our bustling twenty-first-century cities people are clamouring for it. And even though vitamin D supplements can be used to manage rickets, and antibiotics given to fight off stubborn infections, there are other reasons why our access to sunlight is more important than ever.

* * *

Less than three hours by car from Hanna and Ben's house in rural Pennsylvania lies New York: the city that never sleeps. Arriving in Sonia's dad's truck, fresh from our stay with the Amish, was like being transported into an alternative universe. The blinds in my Airbnb room in Lower Manhattan were broken, but the light made little difference to my sleep, as I was kept awake by the constant roar of the city: first late-night revellers; then the sound of garbage trucks and trash collectors; then the growing rumble of cars and pedestrians as the working day began.

New York is one of the most densely populated areas in the world, and of its five boroughs Manhattan tops the tables – although its population has fallen since the early twentieth century, when whole families were crammed into tiny tenement apartments on the Lower East Side and sunlight deprivation was rife.

Such was the demand for land that many developers tried to maximise the space they owned by building upwards. By now, the link between sunlight and diseases such as tuberculosis and rickets had filtered into the public consciousness, and people were talking about a 'right to light' – similar to the 'right to light' in contemporary English law, itself based on the Ancient Lights law, which was motivated by the simple desire to have enough light by which to see in your own home. In the UK, this law enables homeowners to block developments that will obstruct their daylight, provided they have enjoyed access to it from across a neighbour's land for twenty years or more.

The growing clamour in New York prompted the authorities to introduce zoning regulations in 1916, which specified that after a certain height, developers would have to 'set back' the upper floors of buildings, leading to the classic 'wedding cake' design of many Manhattan skyscrapers.

These issues have recently returned to the fore, as Manhattan's population begins to expand again. The New York city planning department estimates that by 2030, Manhattan will house between 220,000 and 290,000 more people – roughly one new neighbour for every six current residents. Unsurprisingly, this influx is generating a demand for growth in pockets of the city, which still – arguably – have room to be developed.

Like many American cities, Manhattan is laid out on a grid system, which is perfectly rectangular – except for Broadway, which seems to snake through the neatly ordered boxes of concrete as it chooses. While uptown is generally considered to lie to the north, and downtown to the south, the grid is actually rotated 30 degrees east of north. This means that on two days a year, 5 December and 8 January, the rising sun aligns precisely with the street grid, flooding both the north and south sides of each cross street with light. And on 28 May and 11 July, the towering pillars of glass and concrete neatly frame the setting sun – a phenomenon dubbed 'Manhattanhenge' that draws thousands of tourists and office workers out on to the streets to observe it.

The shimmering towers of Manhattan are impressive to behold, reflecting sun and sky in their substance, as many do. But on the ground, it's a different story. As the city grows upwards, New Yorkers are progressively being deprived of their lunchtime dose of sunshine as their public outdoor spaces are plunged into shade.

Home to iconic skyscrapers such as the Chrysler Building, Rockefeller Center, and United Nations Headquarters, east Midtown is a densely packed district of Manhattan that will seem familiar to anyone with access to a TV set. Yet the city planning department believes that there's still room to grow – particularly at its outer edges, where the buildings are just eight to ten storeys high, and the sidewalks are planted with trees.

Here, next to a small synagogue, and opposite two unpretentious Thai and Japanese restaurants, is the entrance to the 'vest-pocket' Greenacre Park – a space so tiny, that on my first attempt to find it, I overshot and almost missed it entirely.

The park was opened for the people of New York City by the late philanthropist Abby Rockefeller Mauzé, in 1971, 'in the hope that they will find here some moments of serenity in this busy world'. There's a certain irony in the fact that the inherited wealth that funded this haven of sunlight and serenity came from Abby's grandfather, John D. Rockefeller Senior, whose fortune was built on refining oil to kerosene, the market for which was fuelled by the growing demand for indoor living.

Greenacre Park is no bigger than a tennis court, and is entered through a wooden gateway, which supports a pergola that extends around the left-hand side of the park to a raised area, where people sit around chatting and eating their lunch. In her old age, Mrs Rockefeller used to sit here herself, reading, chain-smoking and admiring the enormous waterfall that gushes out of the leaf-covered back wall into the rectangular pool below. The other unusual feature is the grove of spectacular honey locust trees planted in the park's central area. These slender, reddish-brown-trunked trees form a canopy of delicate, fern-like fronds that filter and dapple the sunlight, scattering it into a constantly moving mosaic of softly dancing light and shadow. Combined with the waterfall, it's like being transported into a Hawaiian glade, right in the middle of New York City.

Next to a small café by the entrance, I chance upon Charles 'Charlie' Weston, an African American man in a brown park-keeper's uniform, who helped build the park. If I want to know how it might look if the proposed rezoning goes ahead, he suggests, I should head over to Paley Park, another pocket park located between Madison and Fifth Avenue. I take his advice, and discover an almost identical space, complete with waterfall and honey locusts – yet here the shade from the encroaching skyscrapers has robbed the trees of their mid-level growth, taking much of the park's character along with it.

'Before the skyscrapers, there was a lot of sun; very beautiful trees,' says Tony Harris, another long-standing park-keeper, who has worked here for three decades. 'Nowadays, we still get the sun, but it comes and goes.' I ask if it's made a difference to the number of visitors, and he cracks a broad smile: 'Of course not; it's Paley Park: best-kept little park in the whole of New York.'

Others disagree: a shady park soon becomes a disused park – particularly during the colder months when the lack of sunshine makes it too unpleasant to linger outdoors. In a place like New York, where the price of land is at such a premium, it's hard to justify maintaining a disused park, which puts these outdoor spaces at risk. The 'Fight for Light' campaign centring around Greenacre Park says a lot about people's inherent thirst for sunlight – although, at the time of writing, it had fallen on deaf ears, and the Midtown rezoning plans were expected to go ahead.

Similar battles are raging elsewhere. In London, Roman Abramovich's plans to build a new £1 billion stadium for Chelsea Football Club have clashed with the desire of local families to maintain access to sunlight in their homes and gardens.[13] Even in sweltering Delhi, where the sun shines on 350 days of the year, a relaxation of the height restrictions on new buildings has prompted concerns about neighbouring buildings being cast into shadow. It's already an issue in Mumbai, where a recent report by the consultancy firm Environment Policy and Research India recommended at least two hours of 'uninterrupted sunlight' for buildings each day.[14]

The discovery that city dwellers' access to sunlight is crucial for their physical health was hard-won. Now, as ever more of us pour into already crowded cities, we risk forgetting those lessons. Parks and other public outdoor spaces shouldn't be a luxury. A recent review of published evidence by the World Health Organization concluded that access to urban green space is beneficial to people's mental health, reduces the toll from cardiovascular disease and type 2 diabetes, and improves pregnancy outcomes. And for children and young people, there's another reason why spending more time outdoors is good for them – and it has to do with the shape of their eyeballs.

* * *

Ian Morgan lays claim to having once been the world's foremost expert on the chicken retina. Most of all, he was interested in how the eye makes the switch from low-light to high-light vision – a process that involves the signalling molecule dopamine. 'If you tell that to people at a dinner party they fall asleep around you,' says Morgan, in a strong Australian accent. Tell them you're working on a cure for myopia, however, and they start to sit up and take notice – particularly when you mention that your work may change the fate of a billion children in China.

East Asia is grappling with an epidemic of myopia (short-sightedness), a condition that, like rickets, begins in childhood. In many urban areas – including Guangzhou – prevalence often exceeds 90 per cent, whereas sixty years ago, just 10 to 20 per cent of the Chinese population was short-sighted. This is very different from Morgan's native Australia, where just 9.7 per cent of Caucasian children are short-sighted.

Guangzhou, in South China, is at the heart of one of the most populous, built-up areas on the planet. It's also home to China's largest eye hospital, where Morgan is a visiting fellow – yet even here they are struggling to keep up with demand for their services. On some days, you can't pass through the corridors for all the patients.

Yet they are the lucky ones: in some rural areas, a misconception that glasses will harm children's eyesight means that many are left untreated. Because they can't see the blackboard, they fall behind with their schoolwork.

So few people are short-sighted in Australia that Morgan wasn't even entirely sure what myopia was during the early part of his career. However, every now and then an academic paper on the subject would cross his radar, so one day he decided to start reading.

This taught him two things: 1) although many textbooks claim that myopia is a genetic condition, its prevalence is rising far faster than can be explained by natural selection; and 2) myopia isn't simply a case of needing glasses: it is also a leading cause of adult-onset blindness.

When Morgan heard about the soaring levels of myopia in East Asia, he spotted an opportunity to make a genuine difference to people's lives, and set out to investigate. His first task was finding out just how prevalent the disorder was in Australia compared to neighbouring Asian countries. His results were startling.[15] By the age of seven, the prevalence of myopia in Australia was 1 per cent; in Singapore it was 30 per cent. Wondering if genetics might have something to do with it, Morgan then looked at myopia prevalence among children of Chinese ancestry who had been raised in Australia: it was just 3 per cent.

'The only factor we could think of that differed, was the amount of time that kids reported being outdoors,' Morgan says. Further studies revealed that, whereas Australian children spend some four to five hours outdoors per day, in Singapore it was more like 30 minutes.

Morgan's theory that outdoor light might be protective was strengthened by a series of animal experiments by other labs. Raising baby chickens in low light was found to significantly increase their chances of becoming myopic, and a separate study found that housing chicks in the equivalent of outside light protected them from developing an experimental form of myopia.

Myopia is caused by the eyeballs growing too long, resulting in the light from distant objects focusing a short distance in front of the retina, rather than directly onto it. In severe cases, the inner parts of the eye stretch and thin, resulting in complications such as cataracts, retinal detachment, glaucoma and blindness.

The current best guess is that light stimulates the release of dopamine in the retina, which blocks the elongation of the eye during development (unfortunately, bright light doesn't seem to reverse myopia in adults). Retinal dopamine is regulated by a circadian clock and usually ramps up during the day, enabling the eye to switch from night- to daytime vision. The theory is that, in the absence of bright daytime light, this rhythm becomes disrupted, resulting in retarded growth. Further studies have revealed that intermittent exposure to bright light – as one would naturally get if one spent lots of time outdoors on this rotating planet – is even more protective against experimentally induced myopia.

Ironically, given the impact that myopia can have on children's education, it is the very desire to see their children excel at school that is contributing to the problem in East Asia. The combination of intensive schooling and a lifestyle that actively discourages children from playing outside, is depriving them of daylight – an essential ingredient of healthy eye development. 'Children don't go outside at recess time, because they're told it is bad for their skin; that girls will never get a husband if they have dark skin,' says Morgan. 'In Australia it's a punishment if you don't go outside.'[16]

However, myopia isn't an issue only for East Asia. In both the UK and US, myopia rates have doubled since the 1960s, and continue to rise. In Western Europe, some 56 per cent of people are expected to be short-sighted by 2050; in North America it's projected to be 58 per cent. And even outdoor-loving Aussies aren't immune from the trend towards increased indoor living and time spent in front of electronic screens: based on current trends, some 55 per cent of Australasians are projected to be myopic by 2050.

Once myopia starts, it generally progresses until late adolescence, so if you can delay its onset by even a few years the hope is that this may drastically reduce the numbers suffering from severe myopia, with its associated risks.

The solution may be relatively simple. In 2009, Morgan launched an ambitious trial to put his theory about the protective effects of outdoor light to the test in Guangzhou. At six randomly selected schools, the six- and seven-year-old children would have a compulsory 40-minute outdoor class bolted on to the end of each school day. The children were also sent home with activity packs containing umbrellas, water bottles and hats with an outdoor activity logo; and they were rewarded if they completed a diary of weekend outdoor activities. Children at six additional schools carried on as normal, serving as controls. Three years later, Morgan and his colleagues compared rates of myopia between the two sets of schools: in the schools with the outdoor intervention, 30 per cent of children had developed myopia, whereas in the control schools it was 40 per cent. [17]

That may not sound like much, but the children of Guangzhou were only getting an extra 40 minutes of daylight per day, five days a week during term time – and from a very low starting point. Also, barely any families took up the challenge of venturing outdoors more frequently at the weekends. 'Our hypothesis would be that you need to be aiming for four or five hours of outdoor time each day, like Australian children have,' Morgan says. Another American study found that spending ten to fourteen hours per week engaged in 'outdoor and sport activities' was associated with approximately half the risk of developing myopia, compared to children doing less than five hours per week.

In Taiwan, schools have tried being more forceful. In 2010, the Taiwanese government launched an initiative called Recess Outside Classroom, recommending that elementary schools (catering for seven- to eleven-year-olds) turn children outside during their recess periods, which total 80 minutes each day. 'They made the kids get outside, turned off the classroom lights, and locked the classroom doors,' Morgan says. After one year, they reported that myopia incidence had halved among schools that adopted the programme.

Such assertive methods are unlikely to work everywhere. Neither is it necessarily healthy to send children to sit outside in direct sunlight: suffering sunburn during childhood or adolescence more than doubles an individual's chances of developing the potentially deadly skin cancer melanoma in later life. However, avoiding direct sunlight altogether can create other issues. Rickets is still a problem in many East Asian countries – and is making a comeback in Western cities, such as London, due to a combination of malnutrition and indoor living.

Researchers are also waking up to the possibility that vitamin D has other important effects on our health – even before we are born – and that the sun may affect our biology in other, unexpected, ways. That mystery factor that so fuelled the popularity of sun cures during the first half of the twentieth century is becoming less mysterious by the day. Sunlight may not be the 'elixir vitae' that Victor Dane proposed – and there's no doubting that it's harmful in large amounts – but the sun's influence over our biology runs deep, and there are some recesses that are only beginning to be explored.

5

Protection Factor

WHICH STAR SIGN are you? It may not be a question that you'd expect to hear from a scientist, but your month of birth really does appear to have an influence on your life – or, at least, on your body and your health. If you're born in the summer, for example, you're likely to be taller than average as an adult, whereas autumn babies tend to weigh more at birth and go through puberty earlier. The effects are small (in the case of height, it's a matter of millimetres) but significant. It's almost as if sunlight affects human growth in the same way it does beanstalks and marrows; and, indeed, children also grow faster during spring and summer – as does the hair on your head and the hair of men's beards.[1]

However, the most robust associations between your birth date and later life are to do with your risk of developing particular diseases. As far back as 1929, a Swiss psychologist called Moritz Tramer reported that individuals born in late winter were at greater risk of developing schizophrenia, an association that's been borne out by more recent studies: in the northern hemisphere, people born between February and April are 5 to 10 per cent more likely to develop the disorder, compared with those born at other times of the year.[2] Indeed, the risk is almost twice that associated with having a parent or sibling with schizophrenia. Late spring babies are at greater risk of anorexia and suicide in later life, whereas people with birthdays in autumn are marginally more likely to suffer from panic attacks, and – in men at least – alcoholism.

So, what lies behind all this? Plenty of scientists lay the blame on sunlight – in particular, the amount of sunlight mothers were exposed to during the second half of pregnancy. As we know, sunlight exposure is vital for the production of vitamin D, and vitamin D deficiency is associated with various psychiatric and immune-related disorders.

Various alternative explanations have been put forward for these month-of-birth effects, including temperature, diet and exercise levels – all of which can vary with the seasons. In the case of allergic asthma, individuals born at the end of summer and during early autumn – when there are more house-dust mites around – have a 40 per cent greater risk of developing asthma, which probably relates to when their developing immune systems are first exposed to the allergens that trigger it.[3] There are also seasonal peaks and troughs in the abundance of bacteria and viruses, and how easily they spread. For instance, cold, dry weather causes the fine mist of snot and the viruses we expel when we sneeze to linger in the air for longer, making it more likely that they'll be inhaled by someone else.[4] A mother's exposure to such infections might also influence the way her baby's immune system develops.

Yet the sunlight exposure of the mother remains the strongest suspect in many of these birth-month-related conditions, not least because summer newborns have twice the level of vitamin D in their blood as winter ones do, which demonstrates the magnitude of the difference in sunlight exposure between these seasons. This, or some other factor in sunlight, seems to shape how the baby's body develops, altering their future disease risk.

Sun exposure is not just an issue during pregnancy:

sunlight is implicated in other medical mysteries as well. Various conditions, including type 1 diabetes,[5] asthma, high blood pressure and atherosclerosis are more prevalent among people living at high latitudes – where the days are shorter and the sunlight weaker during winter months – compared with those living nearer the equator. Many of the symptoms of these conditions also tend to improve during the summer months, when there's more sunshine around.

One of the strongest latitude associations is for multiple sclerosis (MS), which is also, interestingly, more prevalent among babies born in spring. MS is an autoimmune disease in which the insulating sheaths around nerves in the brain and spinal cord come under attack. A recent meta-analysis, which combined the results of 321 studies looking at MS prevalence, concluded that for every degree further north or south one travels from the equator, there are an extra 3.97 cases of MS per 100,000 people.[6] MS is also three times more prevalent among people who were exposed to low amounts of sun during youth and adolescence.

If you're looking for a case study on the role of sunlight in MS levels, you could look at its mysterious boom in sunny Iran, a country that, theoretically, you would expect to have relatively low levels of the disease. And historically Iran did have low rates of multiple sclerosis, in common with other Middle Eastern countries. Yet between 1989 and 2006,[7] there was an eight-fold increase in cases to almost six per 100,000 people.[8] Why?

The prime suspect has been a lack of vitamin D, which is increasingly being shown to have other roles in the body besides maintaining healthy bones and teeth. Vitamin D receptors are found in the heart and on pancreatic cells that synthesise insulin,[9] and vitamin D deficiency is associated with both heart disease and type 1 and type 2 diabetes. It influences the development of brain cells, and their signalling and overall health.[10] It is also used by various immune cells to help them fend off attack from foreign invaders and promote wound repair. Of particular relevance to multiple sclerosis, vitamin D also appears to stimulate the development of regulatory immune cells, which can prevent immune reactions from spiralling out of control.

Low vitamin D during pregnancy has been associated with an almost doubling in the risk of the baby developing multiple sclerosis in later life;[11] while young adults with high vitamin D levels are at reduced risk of the disease.

* * *

Sitting relatively close to the equator, and receiving an abundance of sunny days, there should be ample opportunity for Iranians to make enough vitamin D. And until quite recently there was. During the mid-twentieth century, Iran was a country heavily influenced by Western fashions and culture. The last Shah of Iran, Mohammed Reza, who ruled the country from 1941 until 1979, had an inclination for European sports cars, racehorses and American actresses, and he wore Western clothes; while actresses and pop singers were often photographed in miniskirts and bathing suits. All of that changed with the Islamic revolution of 1979. From then on, men were told to dress conservatively, and women were forced to wear long, loose garments, cover their hair and veil their faces – on pain of arrest by the morality police. Skin previously drenched in sunlight was suddenly covered up.

Currently, vitamin D deficiency is high among the general Iranian population, and significantly more prominent among women and children. Data from Harvard School of Public Health also implicates vitamin D in MS, showing that people with lower levels of vitamin D in their blood during the earliest stages of the disease are more likely to develop full-blown symptoms and have a poorer prognosis.[12]

Like circadian rhythms and melatonin, vitamin D has ancient origins; it has been estimated that phytoplankton and zooplankton in our oceans have been producing it for more than 500 million years. The inactive, precursor form of vitamin D is found in most life forms, including these tiny marine plankton: this may explain why the livers of fish – which eat plankton – are such a rich dietary source of vitamin D. In these early organisms, it helps to protect them from the more destructive aspects of the sun's energy, by absorbing some of the UV rays that cause DNA damage.

However, the active form of vitamin D, which is so important to the human skeleton – and the apparatus needed to generate it – are only found in vertebrates.

The trouble is that at latitudes above 37°, which includes anywhere north of San Francisco, Seoul or the Mediterranean Sea, and in most of New Zealand and parts of Chile and Argentina in the southern hemisphere, the amount of vitamin D synthesised during winter months is negligible. In the UK, we can only make it between late March and September, which makes us reliant on reserves of vitamin D built up during sunnier months, as well as dietary vitamin D from sources like oily fish, egg yolk and mushrooms.

The fact that so many of us spend so much of our daytimes indoors has prompted concern that many people at high latitudes aren't storing enough vitamin D to see them through winter – and that their bones, muscles and possibly other tissues are suffering as a result. In 2016, the UK's Scientific Advisory Committee on Nutrition even recommended that all Brits consider taking vitamin D supplements during winter months – primarily to protect their bones. Particularly for elderly people, falls and fall-related fractures are a major source of injury and death, and a major drain on the health service – so this is good advice. Yet the list of other illnesses that have been associated with vitamin D deficiency in recent years is extensive: as well as MS, it includes cardiovascular disease, various autoimmune and inflammatory diseases, infections, and even infertility.

You might conclude, therefore, that taking vitamin D supplements would equal better health. But sadly, for many of these illnesses, this doesn't seem to be the case. This is also true of MS: although lower levels of vitamin D are associated with an enhanced risk of developing the disease, and a more severe disease course, no study has yet shown that vitamin D supplements can improve the symptoms of MS once it has developed.[13]

In late 2017, a review[14] of multiple trials of vitamin D supplements in patients of all ages concluded that the evidence for them preventing or being useful in non-bone-related conditions was sparse, with two exceptions: vitamin D supplements can help to prevent upper respiratory tract infections and the worsening of existing asthma. Taking them is also associated with a longer life expectancy among middle-aged and older people – but mostly for those in hospital or living in an institution, where they don't get outside much. These things are obviously important, but as a panacea for all the health challenges of the twenty-first century, vitamin D supplements are looking decidedly lacklustre.

This isn't necessarily the end of the story for vitamin D. It could be that we haven't yet found the best time to give supplements, or the right dose, or that the trials haven't lasted long enough to detect an effect on our health; also, because many trials have included people with adequate vitamin D levels, this could have concealed the benefits of vitamin D supplements to people with deficiencies. Several large trials are still ongoing, and until the results of them are in, the jury remains out.

However, it's also worth considering whether something else in sunlight is contributing to some of the wider health benefits ascribed to vitamin D, including the reduced risk of developing MS. Vitamin D is clearly good for us, but the level in our bodies is also a strong marker of how much time we've spent in the sun more generally. Popping vitamin D supplements is not the same thing as spending more time outdoors; and if we rely on them to make up for inadequate sunlight exposure, then we may be missing out on something else that sunlight provides.

* * *

Slip! Slop! Slap! As health campaigns go, the Australian Cancer Council's SunSmart slogan, which featured a dancing cartoon seagull advising people to slip on a shirt, slop on the sunscreen and slap on a hat, ranks among the most successful in Australia's history. Launched in 1981, the message seared itself into the collective psyche, and is widely credited with reducing the incidence of basal-cell carcinoma and squamous cell carcinoma – the two most common forms of skin cancer.

In 2007, the slogan was updated to 'slip, slop, slap, seek and slide', to emphasise the importance of also seeking shade and sliding on a pair of wrap-around sunglasses in order to prevent sun damage.

Australia has one of the highest rates of melanoma on the planet. On average, thirty Australians will be diagnosed with it, and three will die from it each day. With all this talk of the beneficial effects of sunlight, it's worth stressing the downside: there is no doubt that exposure to UV light, including sunlight, is responsible for causing skin cancer.

This was recognised as early as 1928, when the fad for UV lamps and sunbaths was approaching its peak. When a British researcher called George Findlay exposed mice to daily irradiation with UV from a mercury arc light, he observed that tumours developed on their skin. Since then, many more studies have strengthened the link between UV exposure and skin cancer, as well as demonstrating that sunscreen reduces the risk of developing it.

The reason is that UV light triggers DNA mutations in our skin cells, causing them to malfunction and start growing abnormally. Increasingly, though, it seems that an additional mechanism may be at work – one that might also explain sunlight's beneficial effects in inflammatory and autoimmune diseases. As ever, sunlight is a double-edged sword: a creator and destroyer of life.

During the 1970s, an American researcher called Margaret Kripke discovered that if she implanted skin cancers into healthy mice, they were rejected, but if she implanted them into mice that had previously been irradiated with UV light, they established themselves and grew.[15] Kripke concluded that UV light must somehow be suppressing the immune system, which could help explain why immune cells – usually so good at detecting and destroying abnormal cells – sometimes fail to detect and reject early skin cancers caused by sun exposure.

In other words, as well as triggering the mutations that cause them in the first place, the reason skin cancers are permitted to grow is because the immune system is dampened by too much sun exposure.

The skin is our largest organ, covering approximately 2 square metres, and weighing some 3.6 kg. According to the *Encyclopaedia Britannica*, skin provides protection and receives sensory stimuli from the external environment. However, it seems that we may have grossly underestimated its function. Recent evidence suggests that our skin is also a vital part of the immune system, conveying information about outside threats to the vast immunological orchestra at its command.

The predominant cell in our outermost layer of skin, the epidermis, is the keratinocyte. As well as producing the structural protein keratin, which renders our skin almost waterproof, keratinocytes are also in constant dialogue with immune cells in nearby lymph nodes, as well as with nerve cells in the skin.

These keratinocytes are covered in receptors that can absorb UV light: they respond to it by sending chemical signals to various immune cells – particularly a subset of 'regulatory' cells, which help to keep the immune system in check – and if the signals are strong enough they will transmit them to the rest of the body, suppressing its immune responses.

Given that we've evolved as daytime creatures on this sunny planet, presumably there's a reason for this immune suppression. One idea is that it's a way of tolerating 'self'; the immune system is a powerful weapon, which, left unchecked, could quickly turn on our own tissues and destroy them, so tolerating 'self' is essential for survival. 'If you break tolerance, essentially the immune system will kill you,' says Scott Byrne, an immunologist at the University of Sydney, who has been investigating this new-found role of UV light. 'By getting sunlight, we are essentially maintaining that tolerogenic environment, which is essential for preventing autoimmune diseases.'[16] On the flip side, if we get too much sunshine, our immune cells also start to tolerate cancers growing in the skin.

Prue Hart, an immunologist at the University of Western Australia, has long been fascinated by the latitude associations for autoimmune diseases such as MS, and she has been disappointed by the results of vitamin D trials, which have failed to show a benefit of supplements in slowing or stopping progression of the disease. However, the discovery that UV light suppresses certain immune responses has inspired her to start investigating UV light as a potential therapy for MS. Already she has shown that by irradiating mice with UV doses roughly equivalent to a brief stint in the midday sun she can prevent them from developing an experimental form of multiple sclerosis, called experimental autoimmune encephalomyelitis (EAE).[17] Now she's working with Byrne to investigate whether UV exposure from phototherapy lamps – more commonly used to treat inflammatory skin conditions, such as psoriasis – could slow, or even prevent, the development of MS in people showing the very first symptoms of the disease.

In a pilot study of twenty people,[18] only seven of the ten who received phototherapy for two months had developed full-blown MS a year later, whereas everyone in the control group had. The UV group also reported feeling less fatigue. Importantly, vitamin D levels remained similar between the two groups, suggesting that this wasn't the reason for the improvement. Although it is still very early days, and larger trials will need to be carried out, these results are a ray of hope for people suffering from autoimmune diseases.

And yet immune suppression doesn't explain everything. It can't, for instance, explain why sunbathers appear to have longer life expectancies, despite the increased risk of cancer that this activity carries.

* * *

Richard Weller started out his career as a 'good' dermatologist, believing that sunlight was dreadfully bad for you, 'because that's what dermatologists say'. He still doesn't dispute that it's a major risk factor for skin cancer. Even when he discovered that the skin can produce nitric oxide – a potent dilator of blood vessels – he assumed that it would be involved in driving the growth of skin cancer rather than being beneficial to our health.

Then he discovered that we stockpile vast quantities of a storable form of nitric oxide in our skin, and that it can be activated by sunlight. That's when the blinkers fell off: 'Hang on,' he thought. 'Maybe that's the reason why people's blood pressure readings are lower in summer than in winter.' [19] Related to this, it may also help explain the greater rates of cardiovascular disease at higher latitudes.

Subsequent experiments confirmed it. If you expose somebody to the equivalent of around 20 minutes of British summer sunlight, they will experience a temporary drop in blood pressure that continues even after they step indoors.[20]

It's not only blood pressure that seems to benefit from the mobilisation of nitric oxide by sunlight. Separate studies have revealed that mice fed a high-fat diet can be protected against the usual weight gain and metabolic dysfunction through regular exposure to UV light.[21] Block nitric oxide production and you block this protective effect. Nitric oxide is also implicated in wound healing, not to mention achieving and maintaining an erection in men. And nitric oxide seems to be another substance to which those regulatory cells that dampen excessive immune reactions respond.

This previously unrecognised interaction between sunlight and our skin could go some way to explaining the perplexing results of the Melanoma in Southern Sweden study.

This study was launched in 1990 to try to gain a better understanding of the risks associated with melanoma and breast cancer. Researchers recruited 29,508 women with no history of cancer, interviewing them about their health and behaviour, and then following them up at regular intervals to see how their health fared.

Among other questions, the women were asked about their sun habits. How often did they sunbathe in summer? Did they sunbathe during the winter? Did they use tanning beds? And did they go abroad to swim and sunbathe? Based on their answers, the women were placed in one of three categories: 'avoidance of sun exposure', 'moderate sun exposure' or 'most active sun exposure'.

Twenty years after the study began, the researchers crunched some of the data, and made some surprising discoveries. The first was that life expectancy among women with active sun exposure habits was one to two years longer than that of sun avoiders. This was even after adjusting for factors such as disposable income, education level, exercise and so on, which might skew the results.

If confirmed, this would put sun avoidance on a par with smoking in terms of its effect on life expectancy, the authors said.[22] Women in the sun-avoidance group had double the death rate, during the study period, of women in the high sun exposure group. Women in the moderate group fell in between.

Controversial as it sounds, the finding fits with other studies that have associated low vitamin D levels with shorter life expectancy. Of course, we now know that sunlight has other effects on our body that could help explain the link, and vitamin D may merely be a marker of overall sun exposure. Alternatively, vitamin D could be having other, unrecognised effects on our biology: ones that prevent early death.

When the Swedish researchers looked at the reason for this reduced life expectancy among sun avoiders, they discovered that it was mostly due to a greater risk of death from cardiovascular disease and other non-cancer-related illnesses, such as type 2 diabetes, autoimmune disease or chronic lung disease.

Another counter-intuitive finding of the study was that active sun-seekers who developed skin cancers other than melanoma had the highest life expectancy of all. Even so, women in the high sun exposure group were more likely to die from cancer than those in other groups, probably because they were living longer. They were also more likely to get skin cancer, including melanoma. Yet if they did get it, their survival rates were higher than those of sun avoiders with the disease.[23]

* * *

All of this leaves health policymakers with a dilemma. Many Australian schools have a 'no hat, no play' policy, to protect children from the sun. This makes sense during the summer months, especially in a country like Australia, where the sun's rays must travel through less atmosphere before hitting the ground, so are stronger. However, similar policies are now finding their way into schools at higher latitudes – including British schools – where the sun is often weaker.

Even Cancer Council Australia, which launched the Slip! Slop! Slap! campaign, has introduced a more nuanced message in recent years in order to reduce the risk of vitamin D deficiency. It is now emphasising the importance of the UV index – a measure of how strong the sun's UV rays are, and therefore when we're at greatest risk of burning – in dictating when sunlight should be avoided. Together with other Australian medical bodies, Cancer Council recommends staying out of the sun when the UV index is 3 or above – even for people who have been diagnosed with vitamin D deficiency – and following the slip, slap, slop, seek and slide message if you are outdoors for more than a few minutes.

However, in more southerly parts of the country during autumn and winter, they are actively encouraging people to head outdoors in the middle of the day with some skin uncovered in order to synthesise vitamin D.

In higher latitude countries such as Britain, the UV index rarely exceeds 3 between October and March, but it can reach 6 on a sunny day in late April, and it may climb as high as 7 or 8 during midsummer. Cancer Research UK recommends thinking about sun protection – especially between 11 a.m. and 3 p.m. – when the UV index is between 3 and 7, and using it at all times when it is 8 and above. An index of 9 or 10 is common during Mediterranean summers – and may even, if rarely, hit 11, which is where the index stops.

Most important of all is to avoid sunburn. If you compare skin cancer rates between outdoor workers and office workers, it's those who spend their working days indoors who are at the greatest risk of deadly melanoma. Outdoor workers are at greater risk of other types of skin cancer, but these are less likely to kill. One reason is that office workers tend to engage in more 'binge-bathing' – that is, heading down to the beach at weekends and overindulging in sunlight, becoming burned in the process: sunburn is a major risk factor for melanoma.

Another possibility is that the difference in outcomes is related to the type of UV rays that people encounter; outdoor workers are regularly exposed to both UVA and UVB rays, whereas office workers may receive relatively high doses of UVA (which can penetrate office windows), but not UVB. Although both sorts of ray play a role in skin cancer, curiously, vitamin D – which is synthesised using UVB light – seems to afford skin cells some protection against DNA damage.

Although few people would currently advocate sun-bathing as a means of avoiding skin cancer, several trials are now investigating whether applying vitamin D directly to the skin could be a way of mitigating some of the harmful effects of sun exposure.

Taken together, these new scientific findings suggest that our transition from predominantly outdoor- to indoor-based lifestyles in recent decades could be having unexpected consequences – including raising our risk of MS, as those studies in Iran have hinted. They also illustrate the pitfalls of trying to replace sunlight, which has shaped our evolution over hundreds of thousands of years, with a single supplement – vitamin D. Although vitamin D is clearly important to many aspects of our health, and supplements are one way of ensuring that those of us living at high latitudes get enough of it during the winter months,[24] they are no substitute for adequate daylight exposure throughout the year (we also need bright daylight to keep our internal clocks synchronised). Too much sunlight is obviously bad for us, but too little also puts our health at risk. The sun should feature in our daily lives, as it has for millennia.

* * *

There's one other thing sunlight does to the skin that warrants mentioning. When sunlight hits the skin, it triggers the production of several molecules that encourage the production of melanin – the pigment that causes skin to tan and affords some protection against sun damage. One of them is ß-endorphin, a substance which triggers the same receptors as opiate drugs like morphine or heroin.

Endorphin release could be another means by which sun exposure reduces the risk of heart disease: by promoting feelings of relaxation, it may combat the negative effects of stress on the heart. Endorphins also activate the reward system, a pathway in the brain that triggers feelings of pleasure in response to specific stimuli – in this case sun exposure – encouraging us to seek them out again. Some regular sunbed users even exhibit physical withdrawal symptoms, similar to those associated with coming off heroin, if they stop tanning.

The release of ß-endorphin in response to sunlight could therefore go some way towards explaining why being in the sun feels so good, and why we so crave it when the sun grows weaker during winter.

6

A Dark Place

AS EARLY AS the second century AD, the celebrated ancient Greek physician Aretaeus of Cappadocia commanded that, 'lethargics are to be laid in the light, and exposed to the rays of the sun (for the disease is gloom)'.[1] *The Yellow Emperor's Classic of Medicine*, a Chinese medical tome estimated to have been written around 300 BC, also describes how the seasons induce changes in all living things, and suggests that during winter – a time of conservation and storage – one should 'retire early and get up with the sunrise ... Desires and mental activity should be kept quiet and subdued, as if keeping a happy secret.'[2] And in his *Treatise on Insanity*, published in 1806, the French physician Philippe Pinel noted a mental deterioration in some of his psychiatric patients 'when the cold weather of December and January set in'.[3]

No more strongly is this felt than at the high latitudes of Scandinavia, where, during winter, daylight dwindles to a mere few hours per day – or vanishes completely. In the northern part of Sweden, winter depression is known as *lappsjuka*, or 'the sickness of the Lapps'. Even the sixth-century historian Jordanes noted the seasonal peaks of cheer and sorrow among the Adogit people who inhabited Scandinavia at that time. 'To have continual light in midsummer for forty days and nights, and likewise no clear light in the winter season ... They are like no other race in their sufferings and blessings,' he wrote.[4]

For the minority of people who suffer from seasonal affective disorder (SAD), and for the very many of us who suffer to some degree from winter blues, winter is literally depressing.[5]

The modern story of SAD as a syndrome dates to the late seventies, when a team of researchers at the National Institute of Mental Health (NIMH) in Maryland, who had been investigating how light affects biological rhythms, were approached by Herb Kern, a short, sixty-three-year-old engineer with a crew cut.

Brimming with energy and enthusiasm, Kern had been keeping detailed records of his bipolar mood swings since 1967, and was convinced that they showed a seasonal pattern, which related to the length and intensity of sunlight. To try to validate his theory, Kern had joined the American Society of Photobiology and had already spoken with several researchers in the field about his mood swings.[6]

Two NIMH investigators, Alfred Lewy and Sanford Markey, had recently published a report on a new method of measuring melatonin levels in human blood plasma: Kern wanted them to test his blood during spring and winter, to see if they could identify biological differences that might account for his changing moods.[7]

Lewy and his colleagues already knew that day length dictated seasonal changes in the biology of certain animals, and that it was the duration of melatonin secretion (that biological beacon of the night) that told their bodies what time of year it was. They had also just demonstrated that melatonin secretion could be suppressed in humans if they were exposed to bright light.

The researchers came back to Kern with a proposition of their own: if the long winter nights really were flooding his system with melatonin and contributing to his depressed mood, then shortening the duration of melatonin secretion, by exposing him to bright light during the morning and late afternoon, should lift him out of it.

Kern agreed to be their guinea pig, and so the following winter – during his annual low – he became the first human to undergo a course of treatment with a light box. Each morning, between 6 a.m. and 9 a.m., he would be bathed in bright white light – just as if he'd flung open the curtains to reveal a clear spring morning. This process was repeated at 4 p.m., when the streets outside were already darkening. After three to four days of this, Kern's mood began to lift; and by the tenth day he was better.

Curious to know how many other people suffered from this strange seasonal illness, another of the researchers, Norman Rosenthal, contacted a reporter at the *Washington Post*, who wrote a story about it. The public response was overwhelming: thousands wrote in, providing a ready cohort of self-selected candidates on which to conduct further experiments with light.

Rosenthal was, himself, sympathetic to their struggles. A native of South Africa, he had arrived in the US in 1976, and quickly experienced a sensation that he'd never encountered before: a draining of energy and a difficulty in achieving all his tasks once the days drew short and dark. As the snow melted, he'd notice his energy returning and wonder what all the fuss had been about those past three months. At the time, the only explanation anyone could give came from his own psychiatric patients, who'd say things like: 'Do you know, everybody in the office has the "Christmas Crunch" and they're all having difficulties.' Rosenthal came up with a new label for this seasonal lethargy and depression: seasonal affective disorder. And so a syndrome was born.

In 1984, he published a paper describing twenty-nine patients – twenty-seven of whom had bipolar disorder – reporting a history of depressive symptoms in winter that disappeared during spring and summer.[8] Again, there was an enormous public response: 'It was as though this something had been there all the time, but it didn't have a diagnosis or label until then,' recalls Anna Wirz-Justice, an emeritus professor of psychiatric neurobiology at the University of Basel, who worked at NIMH at the time.

SAD was formally recognised by the American Psychiatric Association in 1987, although today most psychiatrists regard it as a subclass of general depression or bipolar disorder. In both conditions, some 10 to 20 per cent of patients report a seasonal variation in their symptoms, but the depression associated with SAD does have some unusual characteristics. Whereas people with general depression often lose their appetite and suffer from insomnia, people with SAD often oversleep and overeat (craving carbohydrates is particularly common). Also, the onset of SAD symptoms is usually triggered by shortened daylight exposure as opposed to negative life events.

Statistics on the prevalence of SAD vary depending on the method used to diagnose it, but most studies have used a tool called the Seasonal Pattern Assessment Questionnaire (SPAQ), which assesses seasonal variations in mood, energy, social contact, sleep, appetite and weight. Using these criteria, up to 3 per cent of Europeans, 10 per cent of North Americans and 1 per cent of Asians suffer from SAD. Women seem to be more affected than men, and people who migrate from lower to higher latitudes also seem to be more susceptible.

As you might expect, SAD prevalence varies significantly with latitude. One US study found a prevalence of 9.4 per cent in northerly New Hampshire, 4.7 per cent and 6.3 per cent in New York and Maryland respectively, and just 4 per cent in the balmy southern state of Florida.[9]

Many more people experience a milder form of the

condition called sub-syndromal SAD or the winter blues. In the UK, one in five people claim to experience the winter blues, but only 2 per cent suffer from true SAD.[10] However, estimating the true prevalence is difficult given the subjective nature of symptoms such as mood and lethargy.

That said, there are measurable differences in brain chemistry across the seasons. For instance, brain levels of the mood-regulating neurotransmitter serotonin are highest in summer and lowest in winter in *all* of us – while availability of the amino acid L-tryptophan, which is needed to synthesise serotonin, fluctuates as well.

So, what could be responsible for triggering such changes? There are several theories, none of them definitive. One idea is that people may have retained the same biological mechanism that certain other mammals, like sheep, use to keep track of the seasons. The animals' bodies respond to changes in the duration of melatonin secretion at night. From an evolutionary perspective, it could have made sense to become more lethargic and depressed during the colder months as a means of conserving energy when food would have been less plentiful.

Another theory is that people with SAD are less responsive to light, so that once light levels fall below a certain threshold – particularly if sufferers are spending a lot of time indoors – they struggle to synchronise their circadian clocks with the outside world.

However, the leading theory is the 'phase-shift hypothesis': the idea that later sunrises in winter delay our internal rhythms so that they're no longer in tune with when we go to sleep and wake up. Exposure to artificial light at night could delay us even further. Most people's moods follow a strong circadian rhythm: we tend to wake up grumpy and become more cheerful as the hours pass, then our mood falls again overnight. If this pattern were to become misaligned with the actual time of day, then those lows in mood might occur during the daytime instead. If our bodies were still in 'night mode' when we woke up, we might also feel more tired and sluggish – another common symptom of SAD. Supporting this idea, Lewy has shown that many people with SAD have delayed circadian rhythms. Since bright morning light both advances circadian rhythms and suppresses melatonin, this could explain its antidepressant effect.

Recent insights into how birds and small mammals respond to changes in day length have shed additional light on the matter. According to Daniel Kripke, emeritus professor of psychiatry at the University of California in San Diego, when melatonin strikes the brain's hypothalamus, this alters the synthesis of active thyroid hormone – a substance that regulates all sorts of behaviours and bodily processes, including the production of serotonin, which plays a well-known role in regulating mood.

'When dawn comes later in the winter, the morning end of melatonin secretion by the pineal gland drifts later,' says Kripke. 'From animal studies, it appears that high melatonin just after the time an organism awakens from sleep will strongly suppress the synthesis of active thyroid hormone, and by lowering brain thyroid levels, cause seasonal changes in mood, appetite and energy.'

Quite possibly, many of these factors are implicated, even if the precise relationships haven't yet been fully teased apart. Environmental cues, such as day length and how much sunshine there is, may directly alter the chemistry of the brain, but psychological factors, such as how we respond to these changes and our more general attitudes to winter may play a role as well.

And regardless of what causes winter depression, bright light – particularly when delivered in the early morning – seems to reverse the symptoms.

* * *

While introducing bright light early in the morning has proved a solution to the gloom of winter for some, others have taken a more radical approach.

The inhabitants of Scandinavia are among the most northerly-living people in the world. A tenth of Norwegians live in the Arctic Circle where the sun doesn't rise at all during the dead of winter.[11] Even in relatively low-lying cities such as Copenhagen in Denmark, or Malmö in southern Sweden, the midwinter days are just seven hours long.

Unsurprisingly perhaps, Scandinavia has long been at the forefront of the endeavour to tackle the winter blues – so I travelled there to find out how ordinary people cope with the long winter nights, and to meet some of the experts trying to help them.

The inhabitants of Rjukan, in southern Norway, have a complex relationship with the sun. 'More than anywhere else I've lived, they like to talk about the sun; when it's coming back; if it's a long time since they've seen the sun,' says artist Martin Andersen. 'They're a little obsessed with it.'

Possibly, he speculates, it's because on a clear winter's day, you can see the sunlight shining high up on the north wall of the valley: 'It is very close, but you can't touch it,' he says. As autumn wears on, the light moves higher up the wall each day, like a calendar marking off the dates to the winter solstice. And then as January, February and March progress, the sunlight slowly starts to inch its way back down again, until the town is finally lifted out of the shadows.

Andersen didn't set out to become a sun-chaser. When he moved to Rjukan in August 2002, he was simply looking for a temporary place to settle with his partner and their two-year-old daughter, Sappho; one that was close to his parents' house, and where he could earn some money. He was drawn to the dimensionality of the place: a town with a population of around 3,000, in the cleft between two towering mountains – the first seriously high ground you reach as you travel west from Oslo.

But as summer turned to autumn, Martin found himself pushing his daughter's buggy further and further down the valley each day, chasing the vanishing sunlight: 'I felt it very physically; I didn't want to be in the shade.'

The departing sun left him feeling gloomy and lethargic. It still rose and set each day and provided some daylight – unlike in the far north of Norway, where it is dark for months at a time – but the sun never climbed high enough to be visible or to cast its golden rays down the steep walls of the valley. Rjukan in winter is a grey, flat kind of a place. If only someone could reflect some sunlight down here, Martin thought.

Most people living at temperate latitudes will be familiar with Martin's sense of dismay as autumn's light dwindled and winter set in – and at his yearning for sunlight. There's something about the flat, gloomy greyness of winter that seems to penetrate our skin and dampen our spirits. Few, however, think to build giant mirrors above their town to fix it.

Rjukan was built between 1905 and 1916, after an entrepreneur called Sam Eyde bought the local waterfall and constructed a hydroelectric power plant there. Factories producing artificial fertiliser followed. But the managers of these industries apparently struggled to retain staff because the valley was so glum.

During my visit to the town in early January, I'm awed by the towering Gaustatoppen – arguably the most beautiful mountain in Norway. But down in the valley, despite clear blue skies, the light is flat, and it feels unpleasantly cold. Then, high on the opposite mountainside, I spy a bright flash of light: the *solspeilet* (sun mirrors).

It was a bookkeeper called Oscar Kittilsen who first came up with the idea of erecting large, rotatable mirrors on the northern side of the valley, from where they 'would first collect the sunlight and then spread it like a headlamp beam over the town of Rjukan and its merry inhabitants'.

A month later, on 28 November 1913, a newspaper story described Eyde pushing the same idea, although it would be another hundred years before it was realised. Instead, in 1928, Norsk Hydro erected a cable car as a gift to the townspeople, so that they could get high enough to soak up some sunlight in winter. Rather than bringing the sun to the people, the people would be brought to the sunshine.

Andersen didn't know all this, back in 2002, when he was struck with the same solution of erecting mirrors to counter the gloom. However, after receiving a small grant from the local council to develop the idea, he discovered that he wasn't the first person to consider brightening up the town – even if his fellow visionaries were long dead. He began developing concrete plans, including a mirror mounted in such a way that it turns to keep track of the sun – just as the head of a sunflower does – while reflecting its light down towards Rjukan town square.

The three mirrors, each measuring 17 m², stand proud upon the mountainside above the town. In January, the sun is only high enough to bring light to the square between midday and 2 p.m., but when it's there, the beam is golden and welcoming. Stepping into the sunlight after hours in shade, I am reminded of just how much it shapes our perception of the world. Suddenly, the colours are more vibrant, the ice on the ground sparkles, and shadows appear where none stood before. Within a heartbeat, I feel

transformed into one of those 'merry inhabitants' that Kittilsen imagined.

* * *

Three hundred and fifty miles south of Rjukan, and at
roughly the same latitude as Edinburgh, lies Malmö. In
Sweden, an estimated 8 per cent of people suffer from SAD,
and a further 11 per cent are said to experience the winter
blues. Yet the short days and long nights of the Scandina-
vian winter take their toll on almost everybody.

Following the early experiments with Herb Kern, interest began to grow among psychiatrists about the potential
for bright light as treatment for SAD. Sweden was a particularly vigorous early adopter, although here they went one
step further, dressing patients in white robes and sending
them into communal light rooms.

Malmö psychiatrist Baba Pendse recalls visiting one of
the first light rooms in Stockholm during the late eighties,
with a group of young colleagues: 'After being in there for
some time, we all started to get very lively,' he says. Intrigued
by this response, he began to research light therapy more
deeply, opening his own light therapy clinic in Malmö in
1996.

When I visited Pendse on a drab grey day in January, he
showed me around the clinic and invited me to try a session
for myself.

The *ljusrum* (light room) contains twelve white chairs
and footstools, clustered around a white coffee table, each
draped in a white towel. The table is stacked with white
cups, napkins and sugar cubes. The only non-white object
in the room is a jar of instant coffee granules. It's warm in
the room, and the lights emit a very faint hum.

Approximately one hundred SAD patients use the room

each winter, initially booking in for ten two-hour sessions in the early morning, every weekday for two weeks. Sometimes there is a long waiting list – particularly during late autumn, when SAD symptoms tend to start. Pendse always offers patients the choice of light therapy or drugs for their depression, although, 'unlike antidepressants, with light therapy you get an almost immediate effect', he says. And numerous studies support the idea that light therapy is at least as effective as drug treatment for SAD. There's also mounting evidence that it actively alters the brain's chemistry like a drug.

* * *

Of course, you don't necessarily need a placebo-controlled trial to tell you that being bathed in bright light feels good – and as we saw in the previous chapter, UV from sunlight also triggers endorphins in the skin. It's no coincidence that the beaches of Thailand and other sunny destinations are saturated with Scandinavians from November to March, or that some sunbathers end up suffering from 'tanorexia'.

But beyond light boxes and therapy rooms, Scandinavians may have hit upon another way of recreating that morphine-like high when the sun isn't there – and a powerful defence against the winter blues.

For the past thirty years or so, Lars-Gunnar Bengtsson has visited Malmö's Ribbersborg Kallbadhus almost every day – sometimes twice a day in winter, now that he has retired. 'That's the best time of year to come,' he confides, because the endorphin kick you get jumping from the 85°C heat of the sauna into the 2°C water is much higher. 'Then you really feel it,' he says.

Scandinavians have been taking saunas for at least a thousand years, and archaeologists recently discovered what they think could have been a Bronze Age sauna on the remote Scottish island of Westray in Orkney. Research in rats has revealed a group of serotonin-releasing neurons in the brain, which fire in response to increases in body temperature, and are connected to a mood-regulating area, which may help explain why being in a warm sauna is so pleasurable. Also, like sunlight, saunas have been shown to trigger the release of nitric oxide, which appears to boost cardiovascular health: in a Japanese study of patients with heart failure, regular saunas boosted their hearts' ability to pump blood, and increased the distance they could walk unaided.

Ribbersborg Kallbadhus has combined that intense heat with extreme cold. Its wooden buildings are built on a creaking platform that extends out into the steely-green sea, and both the male and female areas enclose a section of seawater, which is entered down a set of wooden steps.

One regular described the mixed sauna at Ribbersborg as 'like a British pub, but without the alcohol, and where everyone is naked'. Another regular attendee was the so-called 'nude priest', who wrote a column in the local newspaper where he mused on conversations he'd overheard in the sauna. He once noted that it was 'the most democratic place on earth', because when everyone sits in the sauna together naked, they represent only themselves and not their roles in society.

Certainly, the sauna is a sociable place. This is where I first met Lars-Gunnar: he struck up a conversation about the history of the sauna, as I self-consciously sat there – a semi-clad British woman surrounded by naked sweaty men.

People are often less sociable in winter, and I wonder if this sense of community provides an emotional safety net for those feeling a bit low – possibly it's another way of getting through the winter at this latitude. Bengtsson certainly thinks so: 'The regulars come here almost every day: you develop a friendship with them; you talk and hear about their lives and their problems. If somebody misses a day, we wonder where they are; maybe someone takes a bike to go over and see if they're okay.'

The kick you get from plunging from intense heat into icy cold is surely another big draw. Like sunbathing, cold water has been shown to trigger the release of β-endorphin. It also sends out a shot of the fight-or-flight hormone adrenaline, which temporarily blunts pain, causes the heart to race and leaves you feeling exhilarated.

As I open the door from the sauna building and step outside onto the wooden platform, I'm smacked by a blast of Arctic air. The water looks oily in consistency, as if it's been thickened by the cold and is on the verge of forming ice. There's a strong smell of seaweed and brine in the air, not to mention several large and menacing-looking seagulls.

Taking a deep breath, I drop my towel and start walking naked down the wooden steps. The water is painfully cold, so I move quickly, feeling my heart quicken as the water passes my waist, then my breasts, and then my neck. And then I'm out again, feeling a prickling pain move over my skin, which is quickly replaced by numbness, and then the rush: a blissful sensation of being caressed in tiny snow-flakes, and feeling at peace with the world. As soon as it's over, I find I'm craving more.

When I half-jokingly suggest to Bengtsson that he may be addicted to saunas, he nods, with a serious look on his face: 'I chatted with a doctor who works with heroin addicts; he said the same thing is happening in our brains when we go into the sauna – the difference is it's our own endorphins producing these feeling of being happy and at peace with the world.'

* * *

In some parts of Scandinavia, people see no daylight at all for several months of the year. The tilt of the earth is such that, even when areas of the polar region are facing the sun during the daytime, they don't receive any direct sunlight above the horizon. How do they cope with the relentless twilight?

In the case of Tromsø, some 400 km north of the Arctic Circle in Norway, surprisingly well, it seems. Winter in Tromsø is dark – the sun doesn't even rise above the horizon between 21 November and 21 January. Yet despite its high latitude, studies have found no difference between rates of depression in winter and summer.

One suggestion is that this apparent resistance to winter depression at these harsh latitudes is genetic. Iceland similarly seems to buck the trend for SAD: it has a reported prevalence of 3.8 per cent, which is lower than that of many countries further south.[12] And among Canadians of Icelandic descent living in the Canadian region of Manitoba, the prevalence of SAD is approximately half that of non-Icelandic Canadians living in the same place.[13] Even so, Icelanders have a word to describe it: *skammdegisthunglyndi* – the heavy mood of the short days.[14]

An alternative explanation for this apparent resilience in the face of darkness is culture. 'To put it brutally and briefly, it seems like there are two sorts of people who come up here,' says Joar Vittersø, a happiness researcher at the University of Tromsø. 'One group tries to get another kind of work back down south as soon as possible; the other group remains.'

Ane-Marie Hektoen grew up in Lillehammer in southern Norway but moved to Tromsø thirty-three years ago with her husband, who grew up in the north. 'At first, I found the darkness very depressing; I was unprepared for it, and after a few years I needed to get a light box in order to overcome some of the difficulties,' she says. 'But over time, I have changed my attitude to the dark period. People living here see it as a cosy time. In the south the winter is something that you have to plough through, but up here people appreciate the very different kind of light you get at this time of year.'

Stepping into Ane-Marie's house is like being transported into a fairy-tale version of winter. There are few overhead lights, and those that do exist drip with crystals, which bounce the light around. The breakfast table is set with candles, and the interior is furnished in pastel pinks, blues and white, echoing the soft colours of the snow and the winter sky outside. It is the epitome of *kos* or *koselig* – the Norwegian version of *hygge*, the Danish feeling of warmth and cosiness.

The period between 21 November and 21 January in Tromsø is known as the polar night, or dark period, but for at least several hours a day it isn't strictly speaking dark, but more of a soft twilight. The abundance of snow also has the effect of reflecting what light there is upwards, bathing the white-painted wooden houses in a soft, pinkish glow.

Even when true darkness does descend, people stay active: taking the dog for a walk while on skis, or running with a head torch, and countless children keep on sledging and playing in the floodlit playgrounds.

This positive mindset in the face of cold and darkness seems to mark Tromsø out from southern Norway. Kari Leibowitz, a psychologist at Stanford University, spent ten months here in 2014–2015, trying to figure out how people cope, and even thrive, during the cold dark winters. Together with Vittersø, she devised a 'winter mindset questionnaire' to assess people's attitudes to winter in Tromsø, Svalbard and the Oslo area.[15] 'We found that the further north we went, the more positive people's mindsets were,' she says. 'In the south, people didn't like winter nearly as much. But across the board, liking winter was associated with greater life satisfaction and being willing to undertake challenges that lead to greater personal growth.'[16]

* * *

The traditional inhabitants of the far north, the Sami people, also embrace the differences between the seasons, rather than trying to maintain the same pattern of activities and behaviour year-round.

Ken Even Berg is a Sami guide in his late twenties who grew up in the village of Karasjok, some 300 km east of Tromsø, near the border with northern Finland. For most of his life, he has led a traditional, semi-nomadic lifestyle, following the reindeer herd from their winter feeding grounds near Karasjok to their summer grazing near the coast. This takes around ten days in the spring, and ten weeks in autumn, during which time the herders sleep in tents and follow the reindeer on quad bikes.

'For the Sami, it's not so much about light and dark, as when the reindeer are moving,' he says. Because reindeer have no circadian rhythm, this can occur at any time of day or night: 'They move a little bit, then they eat a little bit, then they sleep a little bit,' he says.

And so, the Sami live by the seasons. In spring they often sleep during the daytime, because the snow gets slushy, making it harder for the reindeer to move. At night the ground is rock-hard, and so that's when most travelling occurs.

Summer is a time for chores, such as maintaining

fences, and checking that the new calves are okay. It's also the season when people are at their most sociable and lively. September is the period for rounding up the calves and taking some of them to market; then it's time to begin the migration back east, which gets harder as the days grow shorter. (Autumn also sounds like a fun time for the reindeer, when they feast on hallucinogenic mushrooms, and stagger around like drunk teenagers.)

Winter is a slower time, when the reindeer herders are back in the family home, and the long dark nights make everyone sluggish and less sociable: 'I just don't feel like going out and meeting people in winter, so I stay at home,' Berg says. This fluctuation in seasonal patterns, and annual winter slow-down, is long accepted as part of the Sami's traditional way of life.

So, could adopting a more positive, accepting attitude to winter help others who suffer from SAD or the winter blues? Kelly Rohan, a professor of psychology at the University of Vermont, is convinced it might. She has recently published several trials comparing cognitive behavioural therapy (CBT) to light therapy in the treatment of SAD, and found that the two were roughly comparable during the first year of treatment.[17] Longer term, CBT was even more effective than light.[18] By addressing people's attitudes towards winter, rather than just focusing on their symptoms, CBT breaks the patterns of negative thinking. In the case of SAD, that could be rephrasing thoughts such as: 'I hate winter' to 'I prefer summer to winter', or 'I can't do anything in winter' to 'it's harder for me to do things in winter, but if I plan and put in effort I can', Rohan explains.

'I don't argue that there's a strong physiological component to SAD, and it's certainly tied to the light-dark cycle,' she says. 'But I do argue that the person has some control over how they respond to and cope with that. You can change your thinking and behaviour to feel a bit better at this time of year.'

Finding things to look forward to in winter – be it saunas and ice swimming, or simply cosying down in front of the fire with a good book – could therefore be an effective way of addressing winter blues. And if you can identify enjoyable winter activities that will get you outdoors to reap the alerting, mood-boosting effect of bright daylight, so much the better.

7

Midnight Sun

THE SKY IS POWDER BLUE and the sun magnificent as my mother and I stride through glittering grass and fallen sycamore seeds to Dowth: the fairy mound of darkness. I'm here to experience for myself what it must have been like to be an ancient sun worshipper on the darkest, shortest day of the year.

Older than the Egyptian pyramids, and contemporaneous with the very first phases of Stonehenge, Dowth is one of several passage tombs, mounds and stone circles that were constructed in Ireland's Boyne Valley around 3200 BC. The three largest – Newgrange, Knowth and Dowth – are aligned with sunrise or sunset at key turning points of the year, and decorated with rock art, at least some of which depicts the sun.

The entrances of Newgrange and Dowth are aligned with the midwinter sunrise and sunset, while Knowth is aligned with the spring and autumn equinoxes. This could be a coincidence, except that at Newgrange there's a deliberate opening called a roof-box, which – for 17 minutes on the shortest day of the year – allows light from the rising sun to travel the 19 metres down the low and narrow passageway and penetrate the back chamber, lighting up a sun-like three-spiral design engraved on the rocky back wall.

Gaining access to the annual spectacle at Newgrange is, quite literally, a lottery: each year tens of thousands of visitors compete for just a handful of places in the tomb at midwinter sunrise – and I wasn't one of the lucky winners. However, far fewer people realise that a similar phenomenon takes place at Dowth, and that (for now at least) you can enter the tomb on the afternoon of the winter solstice and observe it.

Unlike Newgrange, there are no tour buses, no glitzy visitor centre marking Dowth's location; only a wooden stile and a small sign on the grassy verge of an Irish country road.

The tumulus rises from the earth like a pregnant belly. Overgrown with gorse and bramble, it seems an unlikely portal for rebirth, which is one theory for why it was created. At the base of the mound, we instinctively turn left, walking clockwise – sun-wise – around it. About halfway round, we find a kerbstone inscribed with circular symbols, picked out with a hammer and stone chisel 5,200 years ago. The seven suns are just as a child would draw them, with rays radiating from a central circle. Five of them are contained within a second circle, creating a wheeled appearance. Possibly, they are depictions of the sun at different times of the year. Others have suggested that they're not suns, but the Pleiades, or Seven Sisters – a bright cluster of stars in the Taurus constellation, visible only during the winter months, which may have been associated with mourning and death.

We continue to circle the base, eventually finding the simple stone entrance to the tomb that is buried at its heart. The mud is churned up around it, and the modern iron gate pushed back, inviting us in. I must stoop double to creep down the narrow passage, stumbling blind into perfect darkness. As I trip on a rounded stone, a gloved hand grabs mine and pulls me leftwards, into a pitch-black chamber at Dowth's heart.

A female voice with a strong Irish accent greets me. It's Clare Tuffy, manager of the Brú na Bóinne visitor centre, whom I'd met at the Newgrange winter solstice celebrations earlier that morning. The chamber we're standing in is circular, and lined with large stone blocks, some of which are engraved with yet more Neolithic art. To the right is a second, smaller chamber, where people with torches are examining some of these symbols. I'm reminded of the numerous caves decorated with Palaeolithic rock art in France and Spain, which our ancestors similarly revered as sacred places. Despite being a refuge of the dead, it's surprisingly warm inside, and feels welcoming, as if we really are inside a womb, rather than a tomb.

At 2 p.m., the event we're waiting for begins. A shaft of light from the passageway begins to penetrate the chamber. The light has a golden quality, and forms a long rectangle on the floor, which grows and slowly creeps backwards as the sun sets lower in the sky. Its progress is impeded slightly by an unkempt cluster of conifers outside, which cast delicate shadows that dance and flicker on the floor. At 3 p.m. – about an hour before sunset – the sunlight hits a series of large stones lining the back wall, illuminating a profusion of marks pecked into them, which are clustered into cup shapes, squiggles and sun-like spirals. One of the stones curves outwards, reflecting the sunbeam into another wedge-shaped recess, where a solar 'wheel' and spiral are carved. There's an awed hush, and we stand in meditative silence, watching the dancing shadows, until at 3.30 p.m., the sunlight begins to retreat from the chamber, slowly plunging it back into darkness.

This phenomenon occurs at Dowth from late November to mid-January, but the strongest illumination occurs on the winter solstice, when the sun is at its lowest ebb. We can only speculate about what our ancestors had in mind when they built this place. Possibly, this sight wasn't intended for the living at all: it was a signal to the dead that it was time to leave their tomb – certainly the journey through the dark tunnel into the light has strong connotations of birth. It also chimes with reports from survivors of near-death experiences, who often describe the presence of a light or the sensation of moving through a corridor or tunnel. Perhaps our ancestors thought that the sun functioned as a guide through the afterlife, or that if the dead followed it, they would be similarly reborn – just as the sun is reborn at this time of year. Certainly, the winter solstice must have been a time of great hope: that light would triumph over darkness, and that life would conquer death.

After the winter, most of us welcome the extended daylight that the spring and summer bring, anticipating an upturn in our spirits and energy levels, as well as warmer weather. Particularly in Scandinavia, the midsummer solstice rivals Christmas for its festivities, as people gather together on midsummer's eve to sing songs, light fires and party through the night. Giant bonfires are lit in many other European countries too: traditionally, midsummer was regarded as a magical time, and these midsummer blazes were believed to ward off evil and to protect crops against disease. In parts of England and France, on midsummer's eve people even used to launch giant burning wheels down hills towards a river – the resemblance of the wheels to the sun unlikely to have been a coincidence. Where they ended up divined the community's fortunes for the coming year.

Midsummer is when the sun's light reaches its zenith; it's when many crops begin to ripen, and plants bear fruit. It's also when many of us feel at our happiest and at our most sociable. But the long summer days bring their own problems: just as too little light is bad for our health, too much of it is problematic as well.

During summer, the light in polar regions is said to be like no other light on earth: 'You get drugged by it, like when you listen to one of your favourite songs. The light there is a mood-enhancing substance,' wrote the American mountaineer Jon Krakauer who summited Antarctica's highest peak, Mount Vinson, during the summer of 2001.[1]

In some cases, the growing light can be deadly. You might assume that suicide rates would be highest in the dead of winter, particularly in those high latitude countries with the shortest days. But although calls to the Samaritans peak around Christmas, suicides, and particularly violent suicides involving hanging, shooting or jumping, peak in May and June in the northern hemisphere, and during November in the southern hemisphere.[2] It is a seasonal pattern that has been demonstrated in numerous studies, across many different countries, from Finland to Japan to Australia. Generally, the higher the latitude of the country, the more suicides there are overall, and the greater the seasonal difference in suicide rates.

One man I interviewed for this book claims that he regularly contemplates throwing himself off the pedestrian bridge over the Mississippi River that he crosses each day, but that these suicidal thoughts reach a crescendo in the spring as he observes the transformation in other people's mood: 'If you are suicidal, and you see the rebirth of life in the spring, the return of birds, the happiness in others as they enjoy the sun and warmth, while you continue to think about suicide ... Those thoughts of "nothing is ever going to change" and "I'll never be happy like everyone else", just get stronger,' he told me.

However, other impulsive acts, such as assaults and murders, also increase as the days grow longer, and these are not likely to be connected to the general improvement in the mood of other people.

One theory is that such actions are provoked by an increase in serotonin levels within the brain as the days grow longer. Although this may sound counter-intuitive because serotonin is usually associated with good mood, SSRI antidepressants, which similarly boost serotonin,[3] are associated with a greater risk of suicide during the first few weeks of taking them. Usually, it takes three to four weeks for their mood-boosting effects to kick in; in the meantime, some people seem to become more physically active and agitated, which could make them more likely to act on suicidal or other aggressive thoughts.

The long, light-filled summer days can also trigger mania in those who are susceptible to it – characterised by excitement, racing thoughts and euphoria – but also irritability, anger, paranoia and delusions. There is even some preliminary evidence that symptoms of mania can be improved by persuading people to remain in a darkened room between 6 p.m. and 8 a.m.

What about those unaffected by depressive illness? It seems likely that these changes in the availability of serotonin and other brain chemicals, brought about by greater exposure to light, affect the healthy too, and these could explain why most of us feel more active, alert and sociable during the brighter months. And as we learned in the last chapter, the greater availability of morning light suppresses any residual melatonin, which may help explain why we feel sharper on summer mornings.

However, the long twilit evenings and bright early mornings can cause an additional problem: insomnia. Human wake times have been shown to track the dawn – at least until the clocks change to daylight saving time, when this natural light tracking system appears to become scrambled. So it is normal to wake up slightly earlier as it gets lighter. Usually, we also go to bed a little earlier too, so that we become more lark-like during summer – although the amount of sleep we get is slightly less.

Even so, too much light infiltrating the bedroom can make it difficult to fall or stay asleep, meaning that the amount of sleep you get is cut short. Preliminary evidence suggests that some people may be more susceptible to the alerting or disruptive effects of light at night than others, including men, and people with blue or green eyes.[4]

The problems of prolonged exposure to bright light are perhaps most striking at the earth's extremes: in Antarctica, sleep problems are so common that workers have their own name for the mild state of delirium they cause: 'Big Eye'.

'You get these incredibly bright, intense days, which in summer last for 24 hours,' says Chris Turney, a British earth scientist, who makes frequent trips to Antarctica and the sub-Antarctic region to collect ice cores for climate research. This constant light can be just as disorientating to a person's perception of time as constant darkness. In the run-up to his death in March 1912, the Antarctic explorer Captain Robert Falcon Scott admitted in his diary to losing track of the days as he and his fellow explorers dragged their sleds through the whiteness.

'The first time I went down there, I remember feeling like I could keep going, and keep going, and I almost didn't want to sleep because my body was so excited,' says Turney. 'Eventually your body just collapses, but I wouldn't say it's a restful sleep, and I often find I get very vivid dreams.'

One of the greatest dangers in this time-free environment is feeling so stimulated by the constant bright light that you simply forget to sleep. On a continent where hypothermia, crevasses and violent storms are a constant risk, tiredness can easily be fatal. 'It doesn't take much for a silly little mistake to blow up and have a big effect, not just on you, but on your other team members,' says Turney.

Working close to the South Pole is also bizarre from

a practical time-keeping perspective. There are no time zones because they all converge there, so the convention is to use the time zone of the country you arrived from. In Turney's case, this was Chile, but just a kilometre away was an American base, which was operating on New Zealand time: Turney's team was working while the Americans were asleep.

The tactics used by Turney and his colleagues in this unusual situation provide some pointers to how we might similarly combat excess light in our night-time environment.

These days, one of the first items on Turney's Antarctic kit-list is an eye mask to block out the light. Studies have shown that wearing an eye mask together with earplugs, when night-time light and noise are an issue, results in more deep- and REM sleep, and in greater melatonin production.[5] Eye masks or blackout blinds are therefore a practical solution to short summer nights, if an imperfect one – because the transition from dark to light when we wake up is so abrupt. There's some evidence that the grogginess, disorientation and confusion that many of us experience when we wake up – called sleep inertia – is reduced if the lights go up more gradually. Therefore, combining blackout blinds with a dawn simulation clock may be a worthwhile strategy.[6]

Turney's team also observes fixed mealtimes, which not only helps to keep their circadian clocks synchronised, but reminds them of what time it is and, in the case of the evening meal, that bedtime is approaching: 'Otherwise people can be up chatting until 2 a.m. or 3 a.m., and then waking again at 5 a.m. or 6 a.m. – and there's a real danger that they don't get a proper rest,' Turney says.

Those who overwinter in Antarctica also suffer from Big Eye. Not only does the absence of daylight cause some of them to free-run, meaning that they become sleepy at unpredictable times, the cold can also make it difficult to fall asleep. Combined with the cabin fever brought about by spending weeks sheltering indoors from the harsh weather, this can fray people's mental health to breaking point.

Although he's never experienced winter in Antarctica, Turney says that there are apocryphal stories of people completely losing the plot. One of them involves several people walking out into the darkness because they couldn't stand it any more; in another, someone got as far as placing their head in a noose, because they just wanted it to end.

The Big Eye experienced by those overwintering in Antarctica highlights the importance of another variable in getting a good night's sleep: room temperature. Your core body temperature naturally falls at night, and this drop reinforces the message that the brain's master clock is receiving from falling light levels, that night is approaching and it's time to initiate the release of melatonin from the pineal gland.

The temperature in our environment is, of course, heavily influenced by the sun and its absence at night. Our ancestors would have felt these changes acutely, but in our modern central-heated homes – just as in extreme environments such as Antarctica – our surroundings can interfere with our ability to dump excess heat at night.

To successfully trigger sleep, your core body temperature needs to drop by about 1°C (2–3°F). Because of this, the UK's Sleep Council recommends keeping a bedroom temperature of 16–18°C (60–65°F). Temperatures above 24°C (75°F) will slow the rate at which heat is lost, while those below 12°C (53°F) will also make it hard to drop off, because the body is doing everything it can to conserve heat.

A warm shower before bed can aid this process – even on hot days – because it tricks the body into releasing excess heat from its core by dilating blood vessels close to the skin.

If the skin remains moist, this process will occur even faster, because as they evaporate the water droplets carry heat with them. The result is that you fall asleep faster and sleep more deeply.[7] This is also why wearing bed socks or sticking a hot-water bottle next to your feet, which are particularly rich in surface blood vessels, can help you to drop off.

* * *

What the experiences of people living and working at these extreme latitudes teach us is that our biology functions best when there is neither too much light nor too much darkness in our environment. What we're looking for, it seems, is a sweet spot where both can flourish; a yin yang that brings harmony to our internal chemistry. It is easy to suggest, a little harder to implement, but worth the effort. And no more so than for sick and frail people, for whom maintaining a strong circadian rhythm and getting a decent night's sleep could spell the difference between life and death.

This doesn't mean that we shouldn't still celebrate those turning points in the year, when daylight is particularly scarce or abundant. On the top of Dowth, I meet four women who invite me to join their picnic of chicken wings and Buckfast; a sweet, caffeine-infused fortified wine with a certain reputation, particularly in Scotland, as 'wreck the hoose juice'. Christmas is just a few days away, and the streets of neighbouring towns are festooned with twinkling lights and decorations. This trip has become an annual pilgrimage for them during this hectic period. At a time when Christmas has become so consumer-driven, they feel that the simple act of sharing a picnic in the low, pale, silvergold light of midwinter is a powerful way of reconnecting with the seasons and putting things back into perspective. One of them is Siobhan Clancy, from Tipperary: 'Sitting here with the sun in my eyes, I feel like there's something in my lizard brain that's saying: "Yes! There's sunlight; you're alive; you're awake; you're getting through winter, and everything's turning again",' she says. 'You don't need to gather fairy lights to get past the darkness if you go outside in the sunlight.' Just make sure that you're getting the balance right.

8

Light Cure

Awake
Wake up to create again
Wake up to remember
Wake up and, again, wake up
Hope gives strength to my alarm clock

MARIA,[1] WHO WROTE THIS POEM, claims to have died and been reborn seven times. Each time she comes out of a depression, she feels like she's starting again from scratch, having to rebuild her relationships, her studio, and her reputation as an artist and teacher. Her depression even led her to attempt suicide in 2008.

Now, though, she's doing well, and the treatment she credits with keeping her depression at bay is unconventional, even counter-intuitive. It involves deliberately depriving her of sleep and bombarding her with bright light, to try to kick-start her sluggish circadian clock.

We've come a long way in the 130-odd years since Finsen established his Medical Light Institute and ushered in a new era of light therapy. Scientists have unpicked many of the mechanisms through which light interacts with our eyes and skin to fine-tune our internal biology. They have established the enormously significant role that circadian rhythms play in preparing our bodies for the various challenges that day and night throws at them. Moreover, they have discovered that misaligned or flattened circadian rhythms – where the differences between the peaks and troughs of various chemicals in the body become less marked – are a common feature of numerous common diseases, contributing both to the progression of the disease, and to how the body recovers from it.

Therefore, if we can strengthen these rhythms and let sunlight back into our lives (while taking care not to burn the skin) it should make a tangible difference to our health and welfare. It's unlikely that strengthening our circadian rhythms is going to cure serious diseases such as dementia or heart disease – but if implementated over the long term, it could reduce our risk of developing them, and if we already have them, reduce the severity of some symptoms.

The medical potential of these discoveries goes far beyond light-related conditions like the winter blues – and it has exciting implications for aiding recovery from serious and hard-to-treat illnesses, such as bipolar disorder, heart disease and dementia. It could also mean that existing drugs for many medical conditions work better, and with fewer side effects. Already, steps are being taken in this direction.

Psychiatry is leading this new field. For the past two decades, Maria's psychiatrist, Francesco Benedetti, has been investigating sleep deprivation, in combination with bright light exposure and lithium, as a means of treating serious depression where drugs have often failed. As a result, psychiatrists in the US, UK and in other European countries are starting to take notice, launching variations of it in their own clinics. The fact that such 'chronotherapy' seems to work, is also shedding new light on the underlying pathology of depression, and the function of circadian rhythms in the brain.

'Sleep deprivation seems to have opposite effects in healthy people compared to those with depression,' says Benedetti, who heads the Psychiatry and Psychobiology Unit at San Raffaele Hospital in Milan. 'If you're healthy and you don't sleep, you'll feel in a bad mood; you won't be able to concentrate; your attention will drop. But if you're depressed it prompts this immediate return to positive mood, and it improves cognitive abilities.'

Just like other organs, the brain shows daily fluctuations in brain cell activity and chemistry, which are thought to be driven by our circadian clock, and by the accumulation of sleep pressure throughout the day. However, in depressed people, both rhythms appear to be disrupted or flat.

Because recovery from depression is associated with a normalisation of these brain rhythms, Benedetti suspects that depression is one consequence of circadian disruption in the brain. And sleep deprivation seems to be a way of jump-starting this cyclic process, speeding up people's recovery.

The first published case of the antidepressant effects of sleep deprivation came from a German physician called Walter Schulte in 1959. Transport infrastructure in Germany was decimated by the war, so when a female teacher received word that her mother was extremely ill, she picked up her bicycle and rode overnight to visit her. The woman, who suffered from bipolar disorder, was depressed when she set off, but arrived well. This report captured the imagination of a young doctor called Burkhard Pflug, who decided to investigate further. By systematically depriving such patients of sleep, he confirmed that spending a single night awake could abruptly jolt people out of depression. The effects were often short-lived, however.

Benedetti became interested in the idea of wake therapy in the early nineties, as a young psychiatrist working in Milan. Prozac had been launched just a few years earlier, hailing a revolution in the treatment of depression. But its effect on some kinds of depression hadn't been properly studied; notably its influence on bipolar disorder, a condition that causes dramatic shifts in mood, from mania – where sufferers become high, over-excited, irritable – to extreme lethargy and depression. Bipolar patients were being excluded from most studies due to the severity of their symptoms.

Benedetti's patients were in desperate need of an alternative to the drugs and treatments on offer. The gauntlet thrown to him by his supervisor was to find a way of making the antidepressant effects of sleep deprivation more enduring.

A handful of American studies had suggested that lithium might prolong the effect of sleep deprivation, so Benedetti and his colleagues retrospectively analysed the responses of those of their own patients who had undergone sleep deprivation and discovered that patients who had been taking lithium were far more likely to show a sustained response than those who hadn't.

More recent studies[2] have shown that lithium ramps up production of a key protein involved in driving the circadian clock in many cells, including those of the brain's master clock – it increases the amplitude of their rhythms. Since even a short nap could undermine the efficacy of the treatment, Benedetti and his team also started searching for new ways of keeping patients awake at night. They learned that bright light was being used to keep pilots alert, so they tried that – discovering that it too prolonged the effects of sleep deprivation. Of course, we now know that bright light can tweak the timing of the brain's master clock as well as boosting activity in emotion-processing areas of the brain more directly. Indeed, aside from its role in SAD, the American Psychiatric Association has concluded that morning light therapy is as effective as antidepressants in treating general depression – even if it is rarely used for this purpose. When light therapy is combined with antidepressants, the effect is even greater.[3]

Benedetti and his colleagues decided to give patients the whole package: sleep deprivation, lithium and light. And the results were promising.

By the late nineties, the clinic was routinely treating patients with this combination, which they called triple-chronotherapy. The sleep deprivations would occur every other night for a week, and the morning bright light expo-sures would continue for a further two weeks – a protocol they still use today. 'We can think of it, not as sleep-depriv-ing people, but of modifying or enlarging the period of the sleep-wake cycle from 24 to 48 hours,' Benedetti says. 'People go to bed every two nights, but when they go to bed they can sleep for as long as they want.'

An exuberant character, whose strongly accented English is accompanied by wildly gesticulating hands, it's hard not to be infected by Benedetti's enthusiasm. But his data speaks for itself: since 1996, the unit has treated close to a thousand patients with bipolar depression – many of whom had failed to respond to antidepressant drugs. Some 70 per cent of these 'drug-resistant' patients responded to triple chronotherapy within the first week, and 55 per cent had a sustained improvement in their depression one month later.

Also, whereas antidepressants can take up to a month to work – if they work at all – and can increase the risk of suicide, the antidepressant effect of chronotherapy includes an immediate and persistent decrease in suicide ideation.

Maria came to Benedetti in 1998, traumatised by an experience on a different psychiatric ward, where she had been physically restrained because of her delusions.

For almost ten years, triple chronotherapy kept her depressions under control, until she was taken off lithium, and relapsed, prompting her suicide attempt. Maria was readmitted to San Raffaele Hospital and underwent triple chronotherapy again, taking a different mood-stabilising drug.

After several attempts, it worked. Now, she success-fully uses it whenever she tips into depression. 'The hardest hours for me are the ones leading up to midnight,' she says. To keep herself awake, she does something physical, such as cleaning. At around midnight, she usually starts to feel more alert, so she may pick up a book and start reading. Although the words might be confusing at first, she will persist. Then, at around 3.30 a.m. or 4 a.m., as the noises of the city start filtering through the walls, Maria may feel the urge to pick up a ball of clay and start mould-ing it. This is what tells her that the treatment has worked, because when she's ill, she can't stand the touch of it against her skin. 'When I am depressed, it feels as if everything is closed away in a box,' she says. 'The day I come back to life, it is as if this box opens up again.'

* * *

Benedetti cautions that wake therapy isn't something that people should try to administer without medical super-vision. Particularly for anyone who has bipolar disorder, there's a risk of it triggering a switch into mania – although in his experience, the risk is smaller than that posed by taking antidepressants. Keeping yourself awake overnight is also difficult, and some patients temporarily slip back into depression or enter a mixed mood state, which can be dangerous. 'I want to be there to speak about it to them when it happens,' Benedetti says.

However, wake therapy is starting to be taken seriously by psychiatrists elsewhere, with countries like Norway leading the way. The drug industry is understandably agnos-tic – after all, you can't patent it. They are coming around to the possibilities presented by a better understanding of the circadian system in mental illness, though. If we can understand what's going wrong with the clock – and how light and/or sleep deprivation fixes it – this could lead to the development of new drugs that replicate, or even enhance these effects.

This interest extends well beyond bipolar disorder. Scientists still have a way to go in terms of unpicking the biological mechanisms underpinning mental illnesses like schizophrenia, major depression, obsessive compulsive disorder and eating disorders. However, they do understand that such conditions are associated with changes in the levels of neurotransmitters, like serotonin and dopamine in the brain – and these neurotransmitters are under the regulation of the circadian clock. What's more, all of these conditions have been associated with disruptions to the circadian clock, or variations in some of the genes that drive it. Episodes are also often preceded by sleep disturbances or circadian misalignment: a hospital near Heathrow Airport admits around a hundred psychiatric patients each year, whose symptoms appear to have been triggered as a direct response to time zone changes following a long-haul flight. There is also mounting evidence that a good sleep routine can improve mental health.

* * *

Of course, circadian disruption doesn't only affect the brain. It can alter people's immunity, as well as bodily functions such as heart rate or digestion – all of which could hinder their health and recovery from illness. Florence Nightingale's observations about the needs of sick people for fresh air and sunlight are at odds with the design of many modern hospital buildings, which are often characterised by small windows and dim indoor lighting that remains switched on day and night. And as we've learned in previous chapters, circadian disruption can be precipitated by exposure to bright light at night, while an absence of bright light during the daytime can cause the daily rhythms in our cells and tissues to flatten out.

Current British guidelines for intensive care units (ICUs) recommend natural daylight in every patient's room, as well as artificial lights that can be dialled up or down. However, even in hospitals that follow this guidance, bedside illuminance during the daytime is similar to that in many offices – and well below the levels found at sunset outdoors.[4] Compounding the problem, certain drugs, including morphine, can alter the timing of circadian clocks,[5] while patients' sleep may be further disrupted by pain, worry or noise. No surprise, then, that hospital patients often have flat circadian rhythms, or rhythms that are out of phase with the external time of day. Some are now starting to ask how seriously this might be impeding their healing and recovery.

The inpatient cardiac unit at Square Hospital in Dhakar, Bangladesh, is located on the tenth floor of a modern building, where patients commonly stay to recover from procedures such as coronary artery bypass surgery. It has views across the city, and each of its bedrooms has windows, although the views and light are obstructed for some patients by privacy screens.

University of Loughborough researchers monitored patients coming in and leaving the unit, and discovered that for every 100 lux increase in illuminance, patients' length of stay was reduced by 7.3 hours. While other studies have shown that having a view also makes a difference, they calculated that light played a more significant role in speeding recovery.[6]

Similarly, a large study of Canadian patients recovering

from heart attacks found that the mortality rate among those recuperating in brighter rooms was 7 per cent, compared to 12 per cent among patients assigned gloomier rooms.[7]

Animal studies are allowing us insight into why this might be. The first few days after a heart attack are crucial in determining how the heart heals, and what the risk is of having another heart attack in the future. This healing response involves immune cells. Studies in groups of mice exposed either to normal or disrupted light-dark cycles after simulated heart attacks showed a significant difference in the number and type of immune cells that rallied to the heart, the amount of scar tissue – and, ultimately, survival rates. Mice whose circadian rhythms were disrupted, as they might be during a hospital stay, were more likely to die from their heart injury.

We know that the cardiovascular system has a strong circadian rhythm: blood pressure is lowest when we're sleeping, but it rises sharply upon waking up; our platelets, small blood fragments that help blood to form clots, are stickier during the day; while the levels of 'fight or flight' hormones like adrenaline, which constrict our blood vessels and make the heart beat faster, are also higher in the daytime. These circadian variations affect the likelihood of suffering a heart attack at various times of day: statistically, you're more likely to have one between 6 a.m. and midday than at any other time.

However, timing may also affect our ability to recover from heart injury. Further studies in mice have revealed differences in the type and number of immune cells that infiltrate injured heart tissue, depending on what time of day the injury occurs.[8] Human studies have also suggested that survival prospects are improved if patients have heart surgery in the afternoon rather than in the morning.[9]

It's not only the cardiovascular system that shows this circadian variation in injury response. Another recent study found that skin cells called fibroblasts, which play a key role in wound healing, may work more efficiently during the day than at night because of fluctuating levels of proteins, which direct the cells towards injured regions. Mice with skin wounds that were inflicted during the night (when they are awake and active) healed faster than those with wounds inflicted during the daytime.[10]

And when the same researchers analysed data from the International Burn Injury Database, they found that people who suffer burns during the night take approximately eleven days longer to heal than those injured during the day.[11] There are numerous other examples of circadian variations in our physiology: viruses find it easier to replicate and spread between cells at night, compared to the daytime; allergic reactions are strongest between 10 p.m. and midnight; while joint pain and stiffness is at its worst in the early morning.

If our circadian rhythms have this powerful an impact on our immune systems, then the disruption of these rhythms – so common in a hospital environment – could impede recovery from serious illness. By the same logic, it's possible that stabilising or strengthening these rhythms, by exposing people to bright light during the day and darkness at night, could enhance their recovery.

Some of the strongest evidence for this comes from studies of premature and low birth weight infants. Although babies are notorious for their fractured sleep, the brain's master clock seems to be in place from about eighteen weeks of gestation. Circadian rhythms mature progressively from that point on, although it's not until around eight weeks after birth that predictable sleep rhythms begin to emerge. The developing foetus isn't exposed to much in the way of bright light, but its burgeoning circadian system may latch on to other cues, such as daily fluctuations in its mother's hormones, heart rate and blood pressure nonetheless. However, if a baby is born prematurely, these signals are lost.

Premature babies are more likely to thrive, it seems, if they are exposed to natural light cycles, consisting of 12 hours of light and 12 hours of darkness. A recent review concluded that such 'cycled light' shortened the amount of time they spent in hospital after birth, compared to infants kept in near darkness or continuous bright light; they also showed a trend towards greater weight gain, reduced eye damage and less time spent crying.[12]

Fewer studies have investigated the impact of light exposure on adult patients, but mounting concern about the effects of hospital lighting on our health is prompting action all the same. The Royal Free Hospital in London is currently installing circadian lighting in its accident and emergency department, while hospitals in several other countries have already introduced it.

Further evidence to support the case for circadian lighting in improving recovery and outcomes for patients is coming from the work of doctors at Glostrup Hospital in Copenhagen. They have been measuring the impact of a circadian lighting system in the hospital's stroke rehabilitation ward, which boosts exposure to bright, blue-white light during the daytime, and both dims and tunes out blue light at night, meaning that patients largely sleep in darkness. When checks or medical procedures are carried out at night, they are performed under amber-coloured light. 'The point is to stabilise their circadian rhythm while they're in hospital, to try and boost their recovery,' says Anders West, a neurologist at the hospital who has been leading the project.

Roughly a third of people experience depression in the immediate aftermath of a stroke, while up to three quarters of them experience fatigue and poor sleep – symptoms that can adversely affect cognitive function, but also their recovery and survival. The data so far suggests that patients exhibit more robust circadian rhythms in response to the circadian lighting system, and show reduced depression and fatigue, compared with those housed in a section of the ward with conventional hospital lighting.[13] West tells me the effect is 'comparable to giving them antidepressants'. Nurses on the ward also say they've noticed a difference – particulary in those patients who also suffer from delerium or dementia. 'They seem to have a better idea about what time of day it is, and I sense that they are calmer,' says Julie Marie Schwarz-Nielsen, a nurse who has worked at the hospital since 2009.

* * *

While it's not currently possible to cure dementia, there is mounting evidence from other studies that quality of life and severity of symptoms can be alleviated by using circadian lighting to reinforce people's body clocks.

Night-waking is a frequent problem for people with dementia (and their carers) – it's a key reason why they often end up in residential care. Not only do they wake up and walk around, which puts them at risk of falling – our balance is under circadian control, and is worse at night, compared with the daytime – but their night-waking is often associated with delirium or confusion as well.

A related problem is sun-downing, where dementia patients become more agitated, aggressive or confused in the late afternoon and early evening. Both phenomena have been associated with disturbed circadian rhythms.

Eus van Someren became interested in the link between the circadian clock and Alzheimer's disease during the mid-nineties. Several studies had indicated that people's circadian rhythms tended to become flattened with age, which is one reason why older people experience shorter and more disrupted sleep. The problem is often worse in institutions like care homes, because their residents are less likely to go outside, and lights may be kept on 24 hours a day, for their own safety.

Van Someren was intrigued by the fact that this circadian flattening – and the problems associated with it – seemed to be particularly pronounced among people with Alzheimer's disease, so he began to investigate further. He discovered that institutionalised patients, especially those who were inactive during the day, were the most affected, and that their sleep issues seemed to get worse as the days grew shorter and lessened as spring and summer approached. More recently, researchers discovered that the weather can also influence night-waking in such populations: it is significantly higher on cloudy days compared to sunny days. In both cases, daylight is the prime suspect.

As people age, the amount of stimulation that the brain's master clock receives diminishes. In part, this is because very elderly people tend to stay indoors, but the lenses of the eye also grow increasingly opaque, and the pupils narrow, which lets less light through. If that wasn't bad enough, people with cataracts often opt to have their natural lenses replaced with artificial ones, which are designed to filter out blue light because it was thought to contribute to macular degeneration – another bane of old age. An unforeseen consequence is a diminished light input to an already under-stimulated master clock.

In 1999, Van Someren, who is based at the Netherlands Institute for Neuroscience in Amsterdam, persuaded the managers of twelve elderly care homes to take part in a clinical trial. Some of the homes would be fitted with additional bright light fixtures – raising the indoor illuminance to the level you'd expect outdoors on an overcast day – and these would remain switched on between 10 a.m. and 6 p.m. every day; the others would continue with their normal indoor lighting. Some residents would also be given melatonin tablets in the evenings, to try to strengthen their circadian rhythms even further.

The lighting didn't cure their dementia, but three and a half years later those residents who were exposed to the brighter daytime lighting showed less cognitive deterioration and fewer symptoms of depression; there was also less deterioration in their ability to get on with everyday life. When bright light was combined with melatonin, the residents also showed less agitation, and slept better.[14]

I went to see how such interventions are working out in practice, visiting the Ceres dementia wing, in Horsens, Denmark. In the orangery, a group of residents sit around playing picture bingo with the staff, natural sunlight supplementing the blue-white lighting overhead. Next to them sits an elderly lady in a blue dress, who is stroking a white and ginger robotic cat, which periodically licks its paw, twitches its ears and moves its head. The atmosphere in the room is calm and cordial: one man is dozing, but most patients are awake and seem relatively engaged.

Jane Troense is a care assistant on the wing. It was she who pushed for the installation of a circadian lighting system, after reading a newspaper article about light therapy for other forms of psychiatric disease. After a little more digging, she discovered that light was also being investigated as a way of improving sleep in patients such as hers.

The main thing she's noticed since the lights were

installed is that the residents seem to be more sociable: 'They look more alert in the daytime and they eat a little more,' she says. They've also dramatically reduced their intake of sleeping tablets and drugs to ease their agitation. Cameras tracking where residents sit during the day have also found that they tend to congregate where the light is brighter.

The lighting hasn't completely alleviated their night-waking – in fact, one severely demented patient had several falls, because the lights were too dim, and on one occasion, walked into a cupboard and couldn't get out. 'She really needed the lights on,' says Troense. Now she has a night light, with the blue part of the spectrum filtered out.

A survey of nursing staff on the dementia wing also revealed that their own levels of distress dropped following the introduction of the lighting: 'That's really important because it relates to how they handle the mental health problems of the residents,' says Katharina Wulff at the University of Oxford, who has been studying the impact of the lighting system. In other words, it may reduce the risk of them venting their own frustrations on the patients.

Even a green parakeet housed in the main corridor of the dementia ward has been affected: before the installation of the new lights, the bird used to chirp and squawk at all hours; now it is peaceful at night.

* * *

It's early days, but as we've seen, circadian realignment and a better understanding of how our internal clocks affect our minds and bodies could markedly improve health outcomes in psychiatric, neonatal and post-operative wards, and in care homes.

However, our increasingly nuanced understanding of

the clocks in our bodies is also being used to make drug treatments work better – and with fewer side effects.

The sheer reach of these clocks is astonishing: almost half of our genes are under their control, and for every major disease or condition investigated so far – including cancer, Alzheimer's disease, type 2 diabetes, coronary artery disease, schizophrenia, obesity and Down's syndrome – genes strongly associated with disease risk have been found to fluctuate according to the time of day.

What's more, over half of the World Health Organization's essential medicines – 250 drugs found in every hospital in the world – hit molecular pathways regulated by internal clocks, which could make them more or less effective depending on when they are taken.[15]

These include the common painkillers aspirin and ibuprofen, as well as drugs for blood pressure, peptic ulcers, asthma and cancer. In many cases, the drugs in question have a half-life of less than six hours, which means that they don't stay in the system long enough to work optimally if they're taken at a sub-optimal time. For instance, the blood pressure drug, valsartan, is 60 per cent more effective when taken in the evening, compared with first thing in the morning. Many of the cholesterol-lowering drugs, statins, are also more effective when taken in the evening.

Such information is rarely communicated outside of academic journals. For example, guidance on the NHS website explains that valsartan can be taken 'at any time of day'. Pharmaceutical industry interest in drug timing also remains small. 'From the point of view of Big Pharma, what they want is a once-a-day white tablet, which lasts a long time and is relatively impervious to the timing of administration,' says David Ray, who investigates circadian rhythms in inflammatory diseases at the University of Manchester.

Yet the idea that the body varies from hour to hour and from season to season is an ancient one. Traditional Chinese medicine describes the vitality of different organs peaking at various times: the lungs between 3 a.m. and 5 a.m.; the heart between 11 a.m. and 1 p.m.; the kidneys between 5 p.m. and 7 p.m., and so on. People's eating, activity, sex and sleep should be timed to coincide with these rhythms, practitioners advise. Similar ideas also pervade Indian Ayurvedic medicine.

Although the explanations for such rhythms sound unscientific to those who subscribe to modern medicine – and it is unlikely that what the ancient Chinese described as heart or liver rhythms bears much resemblance to what we understand about these organs today – it remains interesting that these ancient medics noticed rhythmic fluctuations in our physiology. Certainly, it was what prompted Francis Levi's initial interest in drug timing.

Levi trained as a medical doctor in Paris, but he began looking into traditional Chinese medicine after becoming frustrated by how many of his colleagues treated their patients as objects rather than individuals. Intrigued by the idea that biological rhythms might impact the effects of treatments, he decided to investigate further using modern scientific tools.

Many chemotherapy drugs target rapidly dividing cells, which means that they kill some healthy cells – such as some of the cells lining the gastrointestinal tract, as well as those in the bone marrow – in the process. This explains some of the unpleasant side effects associated with chemotherapy, such as nausea and loss of appetite. However, healthy cells differ from cancer cells in several respects, one of them being that they only divide at certain times of day, whereas these daily rhythms appear to be absent or disrupted in at least some types of cancer.

If he could identify the time windows when healthy cells were dormant but cancer cells dividing, Levi thought, then larger doses of chemotherapy could potentially be given to patients with fewer side effects. It was a radical suggestion, and not all his colleagues approved of it: 'One of the first things I was told was that I should move away from astrology,' says Levi.

Unperturbed, he designed a series of experiments in mice to test whether the toxicity of a new anti-cancer drug – a derivative of the chemotherapy drug anthracycline – varied according to when it was given. He assessed this by recording how much weight the mice lost during treatment, and how the drug affected their white blood cell counts. Sure enough, the drug appeared to be more toxic if it was given during their active nocturnal period, compared with if it was given when they'd usually be asleep.[16] A subsequent trial in women with ovarian cancer confirmed that side effects such as nausea and fatigue could be significantly reduced if the drug was given at 6 a.m. rather than at 6 p.m.[17]

Levi's big break came when his boss obtained access to another new drug, oxaliplatin. Today, oxaliplatin is a blockbuster drug, often used as standard treatment for people with advanced bowel cancer, but in the mid-1980s it had been condemned to the reject pile because it was considered too toxic to be used in patients. Yet Levi's boss was convinced the drug would be effective, if only they could find a way of improving its tolerability. That responsibility fell to Levi.

Again, he started investigating optimal timing for the drug in mice, eventually moving into human studies. His animal experiments suggested that oxaliplatin's toxicity could be lessened if it was given in the middle of the night, when they're most active. To make this work for humans, Levi simply added 12 hours: a crude calculation, but one that seemed to work. A series of randomised controlled trials combining oxaliplatin with the chemotherapy drug fluorouracil found that symptoms such as nausea, loss of appetite and skin reactions were indeed reduced several-fold by timing the doses to coincide with people's circadian rhythms rather than giving them continuously.

To some, the results seemed unbelievable. One of the first patients Levi treated with the oxaliplatin regimen – a man with advanced colorectal cancer – even called him up to complain about being given a sham treatment. 'He said "you're fooling me around; you must have given me a placebo drug, because I have absolutely no symptoms",' Levi recalls.

In fact, the man had received a far higher dose of the drug than normal. However, by using a specially designed pump to deliver oxaliplatin in the afternoon, and fluorouracil in the early morning, the usual symptoms associated with cancer treatment were all but obliterated.

Some studies even indicated that chronotherapy might boost the efficacy of the drugs, resulting in enhanced tumour shrinkage and longer survival, compared with when the drugs were given the normal way. A 2012 analysis of oxaliplatin-based chronotherapy showed that it increased median survival by three months compared to conventional drug timing in men. For some reason, women didn't benefit from the altered regime.[18]

Even so, Levi's data was enough to convince the pharmaceutical industry that oxaliplatin was worth a second look. The drug was approved in Europe in 1996, and in the US in 2002.

More recently, Levi and his colleagues have discovered that another chemotherapy drug, irinotecan, is better tolerated in the morning if you're a man, and in the afternoon/

early evening if you're a woman. Radiation therapy for cancer also results in significantly more hair loss when it's used in the morning compared with the afternoon, because hairs are growing faster in the morning.[19]

These time-of-day effects aren't only confined to cancer. For instance, the seasonal flu vaccine was recently discovered to generate four times as many protective antibodies if given between 9 a.m. and 11 a.m., compared with the results when given six hours later.[20] Certain medical tests will also give different results depending on when the measurement is taken, which is why many doctors now take multiple blood pressure readings over the course of 24 hours, before making a diagnosis of high blood pressure.

Since circadian rhythms exist in every tissue investigated so far, it's highly likely that time-of-day effects will show up in other diseases, drugs and treatments, as this area is explored further.

Challenges remain, though. Besides sex differences, there are also inter-individual differences in the precise timing of our rhythms – and there's currently no quick and simple test to confirm the details of an individual's internal clock. Having this information could have benefits beyond just optimising the timing of drugs; it could also tell you if someone's rhythms were weak or disrupted. 'We know that when the circadian rhythm is disturbed, independently of all the other factors that can influence patient survival, our cancer patients do worse,' says Levi. An alternative strategy would be to develop 'delayed action' drugs, which only become biologically active once the hands of the body clock pass a certain time. Teams are currently trying to develop prototypes of such drugs.

There's also a growing interest in creating drugs that could boost the amplitude of our circadian rhythms, rather than relying on light; as well as ones that could shift the timing of our circadian clock more quickly, which would enable people to adapt to shift work, or speed recovery from jet lag.

Florence Nightingale stressed the importance of observing the ebb and flow of sickness, rather than relying on averages, which she felt were often misleading. No doubt she would have been impressed by the recognition of daily rhythms in the body systems that drugs act on, and which enable us to heal. Certainly, she would have encouraged efforts to optimise hospital or care home environments, giving our bodies the best possible shot at getting better. 'It is often thought that medicine is the curative process. It is no such thing,' she wrote in her *Notes on Nursing*. 'Nature alone cures. What nursing has to do is to put the patient in the best condition for nature to act upon him.'[21]

Light, sleep and timing: these are three basic things that have the potential to transform health care.

9

Fine-tuning the Clock

WHILE WE MIGHT ALL ASPIRE to regular bedtimes and regulated light exposure over 24 hours, this isn't, of course,
always feasible: we travel and get jet-lagged, we work shifts.

And no one travels further or experiences a more
unusual relationship with light than astronauts living in
space. So, if we want to learn how to optimise our physical
and mental performance – and reduce the risk of illness or
injury under challenging light and sleep conditions – we
could do worse than to look to NASA.

From space, the sunrise begins as a convex blue streak
in the blackness, marking the border between night and
day. The streak extends outwards and grows wider, turning
whitish at its crest as a yellow puddle forms at its base,
which quickly ignites into a golden, ten-pointed star. The
star grows steadily brighter, until the blue streak looks like
a ring set with the biggest and brightest diamond you've
ever seen, and as this blazing diamond – our sun – moves
higher, the clouds, ice caps and deep-blue ocean of earth
begin to roll into view. This dazzling view of our planet is
short-lived, however: within three quarters of an hour, the
expanding curtain of light has shrunk back and been consumed by a tide of blackness, that spreads across the earth
as if in pursuit of the vanishing sun.

This spectacle plays out sixteen times a day for the
astronauts of the International Space Station (ISS), as they
chase around earth at a speed of 27,000 km per hour in

order to avoid dropping out of the sky. At this speed, they complete a full orbit of the earth every 90 minutes, which means that they'll see a sunrise or sunset every 45 minutes.

The experience becomes more visceral if they step outside the space station and propel themselves across its surface to carry out essential repairs or maintenance. When the sun is in view, the temperature in space is a blistering 121°C, and when it sets, it plummets to −157°C. Although their spacesuits and thermal layers provide some insulation, these extremes are still keenly felt.

For the most part, however, astronauts are enclosed within the confines of the space station where, apart from some small portholes, and the seven large windows of the cupola – the station's viewing deck – the light is dim. Circadian desynchrony is a major problem for ISS astronauts because the light-dark cycle they're exposed to is so unusual. The ISS is darker than most indoor working environments on earth, and the frequent rising and setting of the sun complicates things still further: 'If you go to the cupola right before going to bed, and you look out and see sunrise or sunset, you just got 100,000 lux,' says Smith L. Johnston, a medical officer and flight surgeon based at the Johnson Space Center in Houston. 'You're not going to be able to sleep for two hours because you will just be buzzed.'

On top of this, ISS astronauts often work long hours under high stress to complete their tasks, and must sometimes work 'slam shifts', involving an abrupt change to their sleep schedule to accommodate, for example, a shuttle docking, or to complete a prolonged and technical piece of construction work.

However, NASA's astronauts don't only have to contend with circadian desynchrony when they're in space. Their training plans are laid out eight years in advance, and almost every minute of their time is accounted for. These include frequent trips to Moscow, Cologne and Tokyo for training: 'They can't have two weeks of downtime to recover from jet lag every time they go to Moscow,' says Steven Lockley, a sleep expert at Brigham and Women's Hospital in Boston, Massachusetts.

NASA takes sleep and preventative medicine extremely seriously. Having spent billions building a space station and training the astronauts to fly it – and scarred by the 1986 Challenger shuttle disaster, which killed all seven crew members, and was partially attributed to the culture of excessive working hours and sleep loss – they don't wish to see it undermined by someone falling asleep on the job. 'Nobody gets the kind of work-up our astronauts get each year, apart from pro-athletes, because once they're in, they're such a valuable highly trained commodity that we do everything we can to fly them,' Johnston says.

One area that NASA has focused on since 2016 is an optimised LED lighting system on board the ISS, designed to improve astronauts' sleep and alertness, so that they can rapidly adapt to slam shifts and the unusual conditions of space. Inside each coffin-like crew cabin, you'll find a sleeping bag plus personal items, and these new adjustable, colour-changing lights, which have three settings. Before bed, the astronauts use the 'pre-sleep' mode, which has had the blue part of the light spectrum removed; when they wake up in the morning, they can get an alertness boost and strengthen their circadian rhythm by flipping the switch to a much brighter, blue-enhanced light. This setting is also deployed to help shift the clock forwards or backwards if an astronaut needs to change their sleep schedule because of work demands. During the rest of the day, the lighting on the ISS is blue-white.

Similar principles are applied back down on earth, to help staff in mission control adjust to the night shift: 'Some of them may not be used to working at that time, so when they take a break, every 90 minutes, we let them go in a room, walk on a treadmill, and they get exposed to a bunch of blue light,' Johnston says.

We can learn a lot from NASA's approach to fighting jet lag, as they have turned it into a fine art. Jet lag, and the sleep deprivation it causes, play havoc with concentration, reaction times, mood and mental abilities. Lockley is employed by NASA to draft up separate, jet-lag-countermeasure plans, detailing when astronauts should be seeing light, and when they should be avoiding it; when they should take melatonin or make use of caffeine; and, in some cases, when to eat and exercise.

The general rule is that it takes a day to adjust to every time zone you cross, but Lockley claims that, with appropriate light timing and melatonin administration, it is possible to shift people by two to three hours a day – this would mean getting over the jet lag from a London to New York flight in two days, rather than four or five days.

To do this, you need to ask yourself two questions:

1) What time does your body clock think it is? To figure out when you should be avoiding or actively seeking out bright light – or taking melatonin if you have any (it's not currently available in the UK) – you need to think about what time it is in the country you're leaving behind. This is the time your body clock is currently set at.

2) Do you want to advance or delay your clock? If you're travelling east, you'll want to ADVANCE it, which means becoming more of an early bird. This means avoiding bright light when your body thinks that it's night-time and seeking it out after 6 a.m. in your old time zone.

HOW TO MINIMISE JET LAG

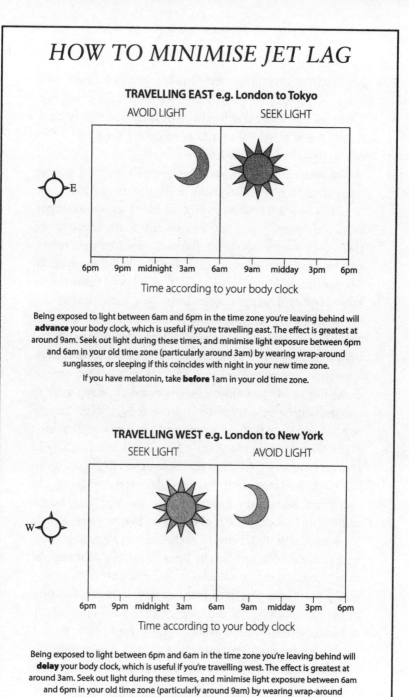

TRAVELLING EAST e.g. London to Tokyo

AVOID LIGHT   SEEK LIGHT

E

6pm   9pm   midnight   3am   6am   9am   midday   3pm   6pm

Time according to your body clock

Being exposed to light between 6am and 6pm in the time zone you're leaving behind will **advance** your body clock, which is useful if you're travelling east. The effect is greatest at around 9am. Seek out light during these times, and minimise light exposure between 6pm and 6am in your old time zone (particularly around 3am) by wearing wrap-around sunglasses, or sleeping if this coincides with night in your new time zone.

If you have melatonin, take **before** 1am in your old time zone.

TRAVELLING WEST e.g. London to New York

SEEK LIGHT   AVOID LIGHT

W

6pm   9pm   midnight   3am   6am   9am   midday   3pm   6pm

Time according to your body clock

Being exposed to light between 6pm and 6am in the time zone you're leaving behind will **delay** your body clock, which is useful if you're travelling west. The effect is greatest at around 3am. Seek out light during these times, and minimise light exposure between 6am and 6pm in your old time zone (particularly around 9am) by wearing wrap-around sunglasses, or sleeping if this coincides with night in your new time zone.

If you have melatonin, take **after** 1am in your old time zone.

If you're travelling west, you'll want to DELAY your body clock, which means becoming more of a night owl. This means seeking out bright light when your body clock thinks that it's night-time and avoiding it after 6 a.m. in the country you're leaving behind.

In both cases, you should be going to bed and waking up at your preferred sleep time in the new time zone.

Let's use a London to Tokyo flight as an example. During the winter months, Tokyo is nine hours ahead of the UK, which means advancing the body clock by nine hours, and becoming an extreme early bird by British standards. Say your flight leaves at 7 p.m. (UK time), and takes twelve hours, you will be arriving in Japan at 4 p.m. local time – but what matters to your body clock, is that it's 7 a.m. UK time. To advance the clock, you will therefore need to avoid light exposure for almost the entire flight, and only seek it out at the very end (after 6 a.m. UK time). One way of doing this would be to invest in a pair of dark, wrap-around glasses, which you'd try to wear at the airport, in the run-up to boarding your flight, and certainly once you were on the plane (aircraft cabins are full of artificial light). It would also be wise to try to sleep throughout the flight, using an eye mask. Melatonin can help people overcome jet lag, but only if it's taken at the right time – in this case, you'd take it just before boarding the flight, reinforcing the sleep signal.

From 6 a.m. (UK time), you should remove your glasses, and actively seek out bright light. You will probably be exhausted, but the good news is that you only need to stay awake until your preferred bedtime in your new time zone. In the run-up to bed, avoid bright light, take some melatonin and, hopefully, get a decent night's rest.

With a far-flung destination, such as Japan, you would face an additional problem the next morning, because although your body clock may have advanced by two to three hours, it will still lag behind Japanese time. People are often advised to get outside and start living on their new time zone as soon as they arrive in a new country, but in this case, doing so would be counter-productive. The sun may be up in Tokyo, but your body clock still thinks it's night-time. You want to carry on advancing your clock, but seeing light now will delay it, so you'll need to put your sunglasses back on and avoid light until after lunch. Because of this issue, when travelling very long distances, it makes a lot of sense to start trying to shift your clock a few days ahead of travel, by going to bed progressively earlier in anticipation of travelling east, or progressively later if you're going to be travelling west.

Several apps are coming on to the market that will do these calculations for you. Lockley is even about to launch one himself. Because of scientific disagreement about exactly how long it takes to shift the clock, though, these apps sometimes give slightly conflicting advice. But in all cases, the same principles apply: it's the time zone that your body clock *thinks* it's in that matters.

* * *

Another field at the forefront of jet-lag management is that of elite sports, where frequent travel clashes with the urgent need for peak performance. Rest and sleep are critical to athletes – as attested by well-known sportspeople the world over, notably Roger Federer who is reported to sleep for nine to ten hours per night. But it's not only about feeling sleepy or wide awake at the wrong time of day: jet lag is yet another form of circadian misalignment. If clocks in muscle cells fall out of synchrony with those in the brain, or in tissues that regulate the supply of fuel to the muscles, then their strength, coordination and reaction times can also suffer. Yet professional athletes spend their lives circling the globe to compete.

The American basketball coach Doc Rivers remembers clearly the moment when he finally appreciated the importance of the body clock to his players' performance. Watching his championship-winning team, the Boston Celtics, being hammered by the Phoenix Suns (who had a reputation for sloppy defence), he began wondering if his players were drunk, they were so bad. Rivers became so incensed, he was sent off the court himself, after arguing with the officials.

Yet months earlier, the sleep expert Charles Czeisler had predicted that the Celtics would lose this very match, thanks to a gruelling pre-game schedule that would see them flying straight from a game in Boston to one in Portland on the Pacific coast the following night; and then immediately flying east to Arizona (which is in yet another time zone), to take on the Suns. Czeisler had even warned Rivers that this game would be akin to watching drunken basketball. He should have listened: the Suns eventually beat the Celtics 88–71.[1]

In a game like basketball, millisecond differences in players' speed and reaction times can make a massive difference to the game's outcome.

Since 2016, the sleep expert Cheri Mah has been collaborating with the US cable and satellite sports channel ESPN, on its 'schedule-alert' project, which aims to predict those basketball games that will be won and lost on player exhaustion.

To do so, Mah weighs up individual teams' travel schedules and game density – both of which can affect players' sleep and physical recovery – to come up with the forty-two games providing the biggest competitive disadvantage to one of the teams. The idea is to raise awareness about the importance of sleep to athlete recovery, but gamblers have also been cashing in on some of Mah's predictions.

During the first year of the project, Mah made a correct prediction 69 per cent of the time; and in the case of seventeen 'red alert' games where the competitive disadvantage was judged to be particularly steep, the accuracy of her predictions rose to 76.5 per cent.

The idea that jet lag might affect athletic performance isn't entirely new, though. One of the first studies to examine it began as a spot of lunchtime fun between several University of Massachusetts neurologists in the mid-nineties. Frustrated at the lack of data to illustrate the physical effects of jet lag, they turned to the trove of North American baseball records to examine whether travelling between the East and Pacific coasts (a journey that involves crossing three time zones) had any impact on game outcomes.

Eastward travel is generally considered harder on the body than travelling west, because it requires people to go to bed and wake up earlier (essentially shortening their day), when for most of us, the natural inclination is to stay up later – probably because our body clocks tend to run slightly longer than 24 hours. This tends to make travelling west a little easier to cope with.

The baseball results supported this notion: visiting teams – who tend to be at a natural disadvantage because they are playing away from home – won 44 per cent of games if they had travelled west, but only 37 per cent of games if they had travelled east. Playing in their own time zone was best of all, though: here, visiting teams won 46 per cent of games. Another group recently extended these findings, analysing more than 46,000 baseball games played over twenty years; they showed that the normal 'home-field' advantage was all but eliminated if the home team had recently flown more than two time zones east (and were therefore suffering from jet lag), and the visiting team was from the same time zone.

Mah has also quantified the benefits of longer sleep for athletes: in one recent study she found that by extending baseball players' sleep from 6.3 to 6.9 hours for five nights, they experienced a 122-millisecond improvement in cognitive processing, which – given that a fast ball takes roughly 400 milliseconds to travel from pitcher to hitter – could provide considerably more time to assess the speed and trajectory of the ball's flight.[2] In another study, she found that when college basketball players committed to trying to get ten hours of sleep per night, rather than their usual six to nine hours, they showed a 9 per cent improvement in the accuracy of their free throws, and a 5 per cent boost in sprinting speed.[3] Again, this may not sound like much, but in professional sport, where the margin between winning and losing is so slim, athletes will grasp any competitive advantage.

Sleep and circadian misalignment aside, physical performance also has a circadian rhythm, which closely follows the daily rise and fall in body temperature and alertness. Muscle strength, reaction times, flexibility, speed: all tend to peak in the late afternoon or early evening.

The evening is when most sporting world records have been set; it's also when swimmers swim fastest, and cyclists take longer to pedal to exhaustion. For sports involving more technical skills, such as football, tennis or badminton, performance tends to peak a little earlier, during the afternoon; this is when footballers have been shown to chip, perform keepy-uppies, juggle and volley with the greatest precision. Few athletes are at their best in the morning; although tennis serves tend to be more accurate then, they are faster in the evening.

These circadian differences are less important if you're

exercising for fun or general fitness, although exercising in the early morning could carry a greater risk of injury, so it's worth spending more time warming up at that time of day. However, if you're seeking to gain a competitive advantage, or to set a personal best, time of day really does seem to matter.

It's also an important consideration for athletes competing at an international level, because crossing time zones will alter the timing of their peak performance. Take English rugby players: studies have shown that they are faster and stronger in the evenings, but if the England team flies to New Zealand to compete against the All Blacks, suddenly their performance will be better in the morning – at least until their body clocks adjust. Therefore, many athletes will travel to their destination country a week or so ahead of major competitions in order to give their bodies time to adapt. The clever ones may also tweak their training schedules, to get used to exercising at the time of the competition.

That's if they want to be at their sharpest. The American ski-jump team is rumoured to actively court jet lag: if you're planning to fling yourself off a giant ramp strapped to a pair of skis, it could pay to be a little foggy in the head, to get over the fear. 'You let muscle memory take over,' one US ski jumper commented.[4] 'Sometimes that's better than thinking about what you need to do.'

* * *

Beyond managing the impact of jet lag on whole teams, some sports are beginning to delve into an even more complex and emerging area: chronotyping their athletes.

Although grip strength peaks at 5.30 p.m. on average, it will peak a little earlier for morning types, and a little later for evening types. It's the same with other physical and mental attributes. 'I might tell a coach – these are the players who could perform a bit better during day games, and these are the players who might perform better during night games – although I don't think anyone has ever been stopped from playing a game because of that sort of information,' Mah says. 'I think coaches intuitively know that so-and-so plays terribly in these types of games, because they've looked at them for so long.'

Naturally, NASA is ahead of the game on this and already 'chronotypes' its astronauts – categorising them as early, intermediate, or late types, based on their preferred sleep times – and will sometimes use this information when devising shift-work schedules or deciding when a specific piece of work on board the ISS should take place.

Imagine a world where every employer did this: where night owls could start their day later, to ensure they were properly rested, and team meetings were scheduled for when everyone was likely to be mentally alert and receptive. This seemingly utopian dream may not be so far off for the inhabitants of one sleepy German spa town.

10

Clocks for Society

THE TOURISM BROCHURE for the German spa town of Bad Kissingen features a photograph of a young woman on its cover. Dressed in white shorts and a pink vest, the woman is perched peacefully on a sunny rock overlooking a river, reading a handwritten journal. Emblazoned on the top left of the page is the slogan: *Entdecke die Zeit* – Discover Time.

In the nineteenth century, Bad Kissingen was a fashionable resort for the European aristocracy and bourgeoisie. They came for rest and relaxation; soaking up the classical architecture and fragrant rose gardens, and taking the mineral-rich waters, which (although they may taste of rusty nails) were reputed to cure all manner of ills.

Today, Bad Kissingen is pushing the discovery of a different kind of time; it has rebranded itself as the world's first *Chronocity* – a place where internal time is as important as external time, and sleep is sacrosanct.

In this book, we've looked at the many ways that we – as individuals – can forge a healthier relationship with light. Yet most of us are not free to choose our work or school hours; we have little control over the lighting in our public spaces and external environment; and we are even forced to reprogramme our internal clockwork twice a year because of daylight saving time.

So what changes could society make to better accommodate our body clocks?

Located in the sparsely populated region of Lower Franconia in Bavaria, Bad Kissingen may seem a strange place to start a revolution. But in some ways its geographical location at the heart of Germany – and, indeed, of Europe – makes it the perfect spot to seed an idea that could spread its tendrils far and wide.

This idea germinated in the mind of Michael Wieden, Bad Kissingen's business manager, in 2013. Having followed scientific developments in the field of chronobiology with interest, Wieden realised that, not only could weaving these principles into the town's fabric benefit its residents, it would also make Bad Kissingen stand out from other rival spa towns.

Bad Kissingen has always been about healing and health, he reasoned; so what better way to heal our modern society than by bringing it back into contact with natural light and sleep. Tourists could come and learn about the importance of internal time, then return home and implement the lessons in their everyday lives.

Wieden contacted a chronobiologist called Thomas Kantermann, who was similarly enthused by the idea. As a teenager, Kantermann frequently found himself in the school principal's office, having pushed the boundaries a little too far; now here was a new set of barriers to break down.[1] Kantermann was ready to launch a revolution in the way that society prioritises sleep.

Quickly, the two men began drawing up a manifesto of the things they'd like to change: schools should start later, children educated outdoors where possible, and examinations not conducted in the mornings; businesses be encouraged to offer flexitime, allowing later chronotypes to work and study when they felt at their best; health clinics could pioneer chronotherapies, tailoring drug treatments to patients' internal time; hotels might offer guests variable

meal- and check-out times; and buildings should be modi-
fied to let in more daylight.

In July 2013, Kantermann and Wieden, together with
Bad Kissingen's mayor and town council, and Kantermann's academic colleagues, signed a letter of intent in
which they pledged to promote chronobiology research in
the town, and to make Bad Kissingen the first in the world
to 'realise scientific field studies in a wider context'.[2]

Most controversial of all was their suggestion that Bad
Kissingen should split from the rest of Germany and do
away with daylight saving time (DST) – the practice of
advancing clocks during summer months in order to make
the evening daylight last longer.

* * *

Since 1884, the world has been subdivided into twenty-four
time zones, all referring to the longitudinal meridian that
crosses the Greenwich observatory in London, hence the
name Greenwich Mean Time (GMT). Moreover, roughly
a quarter of the world's population – including most of
the inhabitants of Western Europe, Canada, most of the
US and parts of Australia – also change their clocks twice
a year.[3]

The original idea of DST is attributed to Benjamin
Franklin, who voiced concerns about energy consumption during dark autumn and winter evenings as early as
1784. Even now, lighting accounts for 19 per cent of global
electricity consumption and approximately 6 per cent of
worldwide carbon dioxide emissions, which is yet another
reason to light our homes less during the evenings.

However, it wasn't until 1907 that an Englishman called
William Willett self-published a pamphlet, *The Waste of
Daylight*, and persuaded politicians to argue his ideas

about changing the clocks through the British parliament. Willett believed that aligning work hours closer to sunrise (at least in cities)[4] might encourage people to participate in more outdoor recreation, enhancing their physical well-being, and might keep them out of pubs, reduce industrial energy consumption, and facilitate military training in the evenings.

Sadly, Willett died of influenza a year before his dream was realised: the UK adopted DST in 1916, followed by the US in 1918. Even so, as Winston Churchill noted, Willett 'has the monument he would have wished in the thousands of playing-fields crowded with eager young people every fine evening throughout the summer and one of the finest epitaphs that any man could win: He gave more light to his countrymen.'[5]

There was a significant downside, however – grasped by a fierce opponent of the change, John Milne, who wrote in the *British Medical Journal*, 'for a certain period twice a year, the efficiency of the worker will be somewhat dampened'.[6]

By moving the clocks forwards each spring and backwards each autumn, we are creating another form of social jet lag. One study of American high school students – a population that's already sleep-deprived – suggested that their sleep was curtailed by 32 minutes per night during the week following the spring clock change; they experienced a short-term reduction in reaction speeds and lapses in vigilance as well.[7] Maths and science test scores fall in the week following the start of DST among young adolescents, while one American study found lower annual scores for the SAT tests, which are used to decide university admissions, among US counties that observe DST, compared with those that don't.[8]

In adults, the transition to summer time and the sleep deprivation it causes has been associated with a 6 per cent increase in 'cyberloafing' – spending one's work time on non-work-related websites, such as those purveying photos of cute kittens – on the Monday after the change compared with the week before;[9] as well as an increase in accidental deaths and injuries, including road traffic accidents. US judges have even been found to dole out heftier sentences for the same crimes in the week after the transition. From a health perspective, clock changes have been tied to an elevated risk of heart attacks, strokes, suicide attempts and psychiatric admissions.

Hubertus Hilgers was seventeen when Germany adopted daylight saving time in 1980. Living in the country-side as he did, this meant getting up at 5 a.m., rather than 6 a.m., in order to catch the bus to school, which began at 8 a.m. 'Already, I was finding it difficult to get to sleep at night until after midnight or 1 a.m., and so the next day I really struggled to get out of bed. My school notes got far worse and my grades deteriorated during the half-year that we had summer time, and then improved when we switched back to normal time.'

Hilgers now lives on permanent winter time – 'normal time', as he calls it – in defiance of the rest of German society. Arranging to meet him in the town of Erfurt, a short train ride from Bad Kissingen, involved a layer of mental arithmetic, which he claims keeps the brain sharp, although for me – and doubtless many others he interacts with – it was a pain.

Yet, many find sympathy with his arguments about daylight saving time. In 2015, he launched an online petition, *Beibehaltung der normalzeit* (retaining normal time), which garnered 55,000 online signatures, plus an additional 12,000 handwritten ones – which was enough to get the national newspapers to take an interest. The petition prompted widespread debate in Germany.

* * *

The discussions around Bad Kissingen reignited that debate. If it had rejected daylight saving time, as Kantermann and Wieden campaigned for it to do, Bad Kissingen would have become *the* DST-free town in Europe: 'Every individual and business would have got a big publicity boost from doing that,' says the chronobiologist Till Roenneberg, who also supports the scrapping of the twice annual changeover.

Deliberately putting oneself in such temporal isolation may sound extreme, but there are precedents. For more than half a century, the US state of Arizona has declined to join the rest of the country in its annual spring leap forward to DST – although the Navaho Reservation, which is inside its borders, does. Moreover, the Hopi Reservation, which is inside the Navaho Reservation, follows the rest of Arizona in remaining on winter time – creating a kind of guerrilla doughnut within an already mutinous state. And, until 2005, some counties and cities in Western Indiana observed daylight saving while others did not.

In the end, the Bad Kissingen town council defeated the motion to become DST-free. But even if the town isn't ready to become a poster child for the anti-DST movement, momentum is building elsewhere – in Finland, for instance, where it's light virtually all of the time during summer, but they still suffer the social jet lag caused by the time shift. The EU Commission also recently proposed abolishing DST – although it requires support from the twenty-eight national governments and MEPs before anything changes.[10] Meanwhile, in southern England many would like to see the entire country shifted permanently forward into Central European Time,[11] given that, in Britain, the annual changing of clocks back to winter time means that it gets dark as early as 4 p.m. in December and early January.

This all goes to highlight a central point: our biology is tethered to the sun, yet the clocks society uses to keep time are influenced by a tangled web of political and historical factors.

Take Germany as an example. At its widest point, the country extends across nine degrees of longitude, and the sun takes 4 minutes to pass over each of them, which means that the sun rises 36 minutes earlier at its eastern border compared to its western one. In a country with the same time zone – and the same TV and radio shows, school start times, and work culture – you might expect that everyone would rise at more or less the same time, but Roenneberg has demonstrated that people's chronotype – the normal time they wake and go to sleep each day – is shackled to sunrise. On average, Germans wake up 4 minutes later for every degree of longitude you travel west, meaning that those in the extreme east rise 36 minutes earlier, on average, than those living in the extreme west of the country.[12] A similar pattern[13] has been documented in the US, where those living on the eastern edge of its time zones are more lark-like than those on the western edge, where the sun rises later.

In some cases, this discrepancy between external and internal time is enormous. A key reason why the Spaniards eat dinner so late is because – positioned as they are at the extreme west of the Central European time zone – 10 p.m. is in fact 7.30 p.m. according to their internal time, which is set according to sunrise.

If the UK advanced its clocks to match Germany and France, this would expose people to more light in the evenings, but not the mornings, pushing our internal clocks even later. Yet we'd still be having to get up at the same time each day to go to work or school, potentially making social jet lag even worse. And in mid-December, a switch to CET would mean that the sun would rise in London at 9 a.m., while in Glasgow this would occur at 9.40 a.m. Many office workers would be arriving at their desks while it was still dark outside. The sun would then set at 5 p.m. in London, meaning that the standard nine-to-five worker who didn't go outdoors at lunchtime would spend several months of winter seeing practically no daylight at all.

Russia, which switched to permanent summer time in 2011, performed an abrupt U-turn just three years later, citing the ill health and accidents it caused.[14] Sergei Kalashnikov, the chair of the State Duma Heath Committee, claimed that the switch condemned Russians to increased stress and worsening health, because of having to travel to work or school in pitch darkness. It was also blamed for an increase in morning road accidents. Since 2014, at least some parts of Russia have switched to living on permanent winter time. However, Muscovites now complain of the insomnia brought about by early sunrises during summer, and sales of blackout blinds have soared, which just goes to illustrate the complexity of the issue and how hard it is to get right.

* * *

Yet if we could find a way to cater better to individual groups' circadian needs, perhaps some of the heat would be taken out of the DST debate.

There are few members of society who more obviously find it hard to conform to the early-bird demands of society than teenagers.

Perhaps it's unsurprising, then, that one of Bad Kissingen's most enthusiastic early adopters of the Chronocity idea was the local secondary school, the Jack Steinberger Gymnasium, which caters for around 900 pupils aged ten to eighteen. Here, a group of older students created a questionnaire and canvassed their fellow pupils about whether it would be desirable to start school at 9 a.m., rather than 8 a.m.: the majority said it would. They also chronotyped the entire school and calculated the amount of social jet lag its pupils were suffering from each week. Approximately 40 per cent were experiencing two to four hours of social jet lag,[15] while a further 10 per cent were contending with four to six hours – equivalent to flying from Berlin to Bangkok and back – each week. Although almost three quarters of adults experience an hour or more of social jet lag per week, only a third experience two or more hours.[16]

As we've seen, teenagers are at greater risk of social jet lag because their biological rhythms are naturally shifted later. This makes it harder for them to fall asleep at night, and yet they still must get up in the morning to go to school. To compensate for the sleep deprivation this causes, they then sleep in at weekends.[17]

Teenagers' later chronotype also means that their natural peaks in logical reasoning and alertness occur later than they do in adults. In one study,[18] Canadian researchers compared the cognitive performance of teenagers and adults during the mid-morning, and again, mid-afternoon. The teens' scores improved by 10 per cent in the afternoon, whereas the adults' scores deteriorated by 7 per cent.

One strategy for dealing with this issue is to delay school start times and allow teenagers to sleep for longer in the mornings, as the Jack Steinberger pupils proposed. The US Midwest state of Minnesota was among the first to investigate the benefits of doing so, after the Minnesota Medical Association sent a memo to all school districts urging them to do something to improve adolescent sleep. As a result, several high schools in the Minneapolis suburb of Edina changed their start time from 7.20 a.m. to 8.30 a.m.[19] When researchers from the University of Minnesota investigated the impact of the change, they were surprised to find almost unanimous support for it among students, teachers and parents. Despite parents' fears that they'd use it as an excuse for going to bed later, the teen's bedtimes remained relatively unchanged, but they slept later in the mornings and got more sleep overall. Students said that they felt less tired during the day and thought that their grades had improved, while teachers noticed fewer pupils with their heads down on their desks and reported that the children seemed more engaged and focused. School attendance also improved.[20]

As news of this success began to spread, other schools started changing their hours as well, but no one had done a proper before – and after – study confirming that it made a real difference. Judith Owens is a paediatrician with a particular interest in sleep medicine. When she was called in by her daughter's high school to talk to staff about the potential benefits of starting school 30 minutes later as they had been discussing, she agreed and decided to see if they could produce some more robust evidence. 'Many felt that half an hour wasn't going to do anything – it would just disrupt the school schedule,' Owens recalls. She suggested that they collect data on the students' sleep and mood before and after a three-month trial of the later start.

Owens was pleasantly surprised by the results. Just a 30-minute delay in starting school resulted in pupils getting an extra 45 minutes of sleep per night: 'Anecdotally, they said that they felt so much better getting an extra half-hour of sleep that they were motivated to go to bed earlier and get even more,' Owens says. 'And they could afford to go to bed earlier because they were more efficient at getting their homework done.'

The percentage of students getting less than seven hours of sleep decreased from 34 per cent to just 7 per cent, while the proportion getting at least eight hours rose from 16 per cent to 55 per cent. The kids also rated themselves as less depressed and more motivated to participate in a variety of activities.[21] But the thing that really swung it for Owens was the change in her own daughter, Grace. 'She was like a different person,' she says. 'It was no longer a battle to get her up in the morning; she would be able to eat breakfast; and the start of the day was just pleasant, instead of torture for everybody.'

Owens changed her research focus and became involved in drawing up policy on school start times for the American Academy of Paediatrics, based on the best available evidence. In 2014, they issued a policy statement: starting school before 8.30 a.m. is a key modifiable contributor to insufficient sleep, as well as circadian rhythm disruption, in the adolescent population.[22]

But how late is late enough? Most British schools don't start school until around 8.50 a.m., but one recent study concluded that most eighteen- and nineteen-year-olds don't feel mentally sharp until much later, and therefore possibly shouldn't start their studies until after 11 a.m. In a separate study, the same researchers tested whether moving the start time of an English comprehensive school from 8.50 a.m. to 10 a.m. made any difference to its thirteen- to sixteen-year-old pupils. Rates of absence due to illness fell dramatically following the change: whereas before they had been slightly above the national average, two years after the change they were down to half the national rate. Pupils' school performance also improved: at baseline, things looked grim, with just 34 per cent of students gaining 'good' GCSE grades at age sixteen, compared to 56 per cent nationally. But after the introduction of 10 a.m. starts, this rose to 53 per cent.[23]

Meanwhile, one British sixth form – the independent Hampton Court House school on London's southwest fringe – is starting lessons at 1.30 p.m. and finishing at 7 p.m., enabling students to have 'more independence over how they structure their day'.

Even a 10 a.m. start would be difficult to impose in countries such as the US, where most adults also start work earlier than in Britain. It would require a change of mindset among parents – as well as a more flexible attitude by employers – but the data suggests that it would make a difference to many pupils.

* * *

The tide may be turning in schools, but in the workplace, there's still a way to go. An individual's chronotype is based on his or her sleep behaviour on free days, and a simple way to define it is to look at when the mid-point of sleep occurs: if you fall asleep at midnight on weekends and wake up at 8 a.m., your mid-sleep time would be 4 a.m. Roenneberg has discovered that for 60 per cent of people, the mid-sleep time on free days is between 3.30 a.m. and 5.30 a.m. There are some earlier birds, but a greater proportion of the remainder sleep later than this.

Expecting people to wake at 6.30 a.m. and then to be mentally sharp when they arrive at work at 8 a.m. or 9 a.m. is therefore fighting against nature to some extent. Like physical performance, your mental skills peak and trough at various times throughout the day. Logical reasoning tends to peak between 10 a.m. and noon; problem-solving between noon and 2 p.m.; while mathematical calculations tend to be fastest around 9 p.m.[24] We also experience a post-lunch dip in alertness and concentration between about 2 p.m. and 3 p.m. However, these are averages, so a lark's peak in problem-solving may arrive several hours earlier than a night owl's.

Research into this area is only just beginning, but managers with early-bird tendencies have been found to judge employees who start work later as less conscientious, and to rate their performance lower, compared to those who share such managers' sleep preferences. 'If your boss shows up at 7.30 a.m. and you walk in at 8.30 a.m., he thinks "we've already been working for an hour, and you're going to work an hour less"; he doesn't see that you will stay for an extra three hours after he goes home,' says Stefan Volk, a management researcher at the University of Sydney Business School. 'It also has to do with his mindset: because he is very productive in the morning, he assumes that's the case for everybody, so he feels you are wasting time.'

Not only would a greater appreciation of these individual differences, and allowances for different schedules, help to level the playing field, it could boost workplace productivity, and employees' health and happiness: 'If you are forcing an evening person to show up at 7 a.m., all you have is a grumpy employee who sits there and drinks coffee, procrastinating until 9 a.m. because he simply can't focus,' says Volk.

Such an approach could create a more harmonious and morally sound workplace as well. Sleep deprivation robs glucose from the cerebral cortex, the brain region responsible for self-control. One study found that employees who got less than six hours of sleep per night were more likely to engage in unethical or deviant behaviour, such as falsifying receipts or making hurtful comments to their colleagues.[25] Another found that the timing of unethical behaviour differs according to people's sleep preferences:[26] larks are more prone to behave unethically towards the end of the day, when they are growing tired, whereas night owls are more likely to behave badly in the morning.

Allowing staff to choose their work hours based on

their individual sleep preferences is one solution then. But is it worth the potential disruption it might cause? In a recent study,[27] American researchers piloted a three-month intervention at a global IT firm, which aimed to improve workers' sleep and work–life balance, by helping them to move from a time-based to a more results-based office culture. Rather than judging colleagues on how they spent their time, workers were encouraged to work at whatever time or place they wanted, so long as they achieved specific results, such as delivering finished projects to customers.

Following its introduction, workers' average sleep time increased by 8 minutes per night – adding up to almost an extra hour of sleep over the course of a week. But, perhaps more importantly, the number of times that people reported never or rarely feeling rested upon waking, went down. As one employee who previously had to get up at 4.30 a.m. in order to get an early start at work and avoid the evening rush hour put it: 'If I'm working from home I don't get up until 6.00 or 6.30 and I start working at 7.00 ... I get more sleep than I've had in years.'

* * *

In Bad Kissingen, Wieden's current focus is on establishing a Centre for Chronobiology in the town, which would provide an academic hub for chronobiology research across Europe. The proponents of the Chronocity project hope that this will galvanise the town and lend authority to their efforts: 'If we have a professor of chronobiology based here, who will go out into the community to give lectures and initiate research, it should be easier to open doors to hospitals and businesses and have a greater influence on health,' says the mayor, Kay Blankenburg.

There have been some other victories as well. The

Stadtbad, which oversees tourist and spa facilities in the town, now offers flexible working to its office staff; while Thorn Plöger, the manager of Bad Kissingen's rehabilitation hospital, took the idea so seriously that, at one point, he adjusted all the hospital's clocks, making some a little fast and some a little slow, in order to provoke reflection. 'People are always so stressed about the time,' he explains. 'They would say, "it's 9 o'clock, I must get my medicine", or "I have a date at midday, so I must leave"; I told them, "take it slowly: *entdecke die zeit*".'

Did they respond well, I ask?

'No,' he says, with a mischievous smile, 'they said "you have to change the clocks back".'

Plöger sighs and shakes his head. 'Germany has a problem. People are always watching the clock.' For the Chronocity initiative to work, he explains, it requires a more flexible mindset: one which says, it doesn't matter when you start work, so long as you get the job done. It's about internal time, not what the clock on the wall says.

* * *

In February 2017, Plöger left the clinic to become manager of the Bavarian Rhön, a 480-square-mile area of rolling wild country, dominated by a set of extinct, dome-shaped volcanoes. Having picked up Wieden's mantel and run with it, he is already planning the world's first region that puts internal time at its forefront. At the heart of these efforts will be policies that promote the value of reducing light pollution (and, hopefully, persuade the Rhön's towns and villages of the same), enabling people to sleep more easily and appreciate the spectacular night skies.

Similar seeds are germinating elsewhere, as people wake up to the fact that light does so much more than just enable us to see. The research for this book has taken me to many places, and introduced me to numerous people who, like Wieden, are agitating for a revolution in our attitudes towards light and sleep.

They have convinced me that it is possible to forge a healthier relationship with night and day without returning to a pre-industrial past, where the extremes of light and dark restricted our productivity and made it uncomfortable to live – even difficult to survive – at certain times of year.

We need to spend a greater proportion of our daytimes outdoors, to reap both the biological benefits of sunlight on our skin, and to realign our internal clockwork. However, it would be naïve to suggest that this is achievable for everyone all the time – sometimes we're just too busy to take a walk around the block at lunchtime; it is impractical to walk or cycle to work; or it's just not possible to eat breakfast next to a large east-facing window, bathed in bright morning light. So, we must also strive to find new and innovative ways of brightening our homes and workplaces – as well as dimming our lights in the evenings.

Already, lighting companies are tweaking indoor lighting to make it more like daylight, but in the future lighting may be tailored to the individual: sensors will detect how much light people have been exposed to over the previous 24 hours, possibly in connection with software used to track their sleeping patterns. The lighting at home and at work will then be adjusted to optimise an individual's circadian rhythms and keep them entrained to the sun.

Similarly, new and better ways of keeping tabs on people's internal rhythms will enable drugs to be given when they're most likely to be effective – or maybe become active only once an internal clock-hand passes a certain hour.

And although we haven't yet found a solution to the problem of shift work, it's clear that we should be doing everything we can to minimise circadian misalignment: this means trying to regularise our schedules as well as going to bed early enough to ensure adequate sleep.

We spawned from a revolving planet, itself shaped by starlight. And although we create our own electric stars to light the night, our biology remains tethered to a monarch mightier than them all: our sun.

Epilogue

ONE OF THE FIRST THINGS I did when I began researching this book was to visit Stonehenge at midwinter, when, for a few short hours, visitors are allowed inside the stone circle (usually they must keep their distance). Two years later, I returned as a guest of the Cotswold Order of Druids. Having spent twenty-four months researching the influence of the sun on our bodies and the circularity of our biology, it seemed important to close this spiritual circle as well.

Like the people who constructed the mound at Dowth, the architects of Stonehenge clearly had the midwinter sun in mind when they constructed the iconic stone circle some 4,500 years ago. Sunset on 21 December is when the tallest trilithon frames the weak, golden sun as it sinks below the horizon, ready to be reborn, a little stronger, the next day.

Other evidence supporting the importance of midwinter to these people comes from the nearby settlement of Durrington Walls, with its wooden-posted sister, Woodhenge. This is where the builders of Stonehenge are thought to have lived during its construction. Among the ancient houses, archaeologists have unearthed a pit containing large numbers of pig and cattle bones; the teeth of the pigs revealed that they were all killed at around the age of nine months – which would have coincided with midwinter. Possibly, people gathered from far and wide to feast on them, before processing along the River Avon to the stone circle to observe the sunset.

Times change though, and in order to preserve this ancient landscape, English Heritage now demands that visitors are shuttled to the stones by minibus instead.

We process, three abreast, towards them; a motley collection of felt-cloaked druids and pagans, plus interested 'friends' like me, wearing cream tabards embroidered with a single green acorn. One woman carries a giant basket of mistletoe, others are bearing wooden staffs topped with animal horns or antlers, and there's a lot of Celtic jewellery. I stifle a smile as tourists pull out their smartphones to record this 'traditional' event.

The stones loom before us, larger than I remember, their pitted, lichenous features giving them the appearance of ancient sentinels forming a protective ring around the space we seek to enter. As we process, sun-wise, around them, the only sound is the jingle of morris dancers' bells and a nameless clanging instrument, which its red-robed bearer claims he bought at a Womad music festival.

There's a cobbled-together charm to the proceedings, drawing inspiration from nature and old traditions from the British Isles. As we complete our circle, the chief druidess approaches the two men with wooden staffs, who are blocking our entrance.

'What's your purpose?' they ask.

'To honour the ancestors,' the druidess replies.

The men draw back, and we enter the stone circle, completing another lap of its interior, before stopping and reaching out to hold hands.

The sun is inconspicuous, just like the last time I was here, but it makes its presence felt through the incessant drizzle that falls from the sky. Without the sun there would be no evaporation and no rain, after all.

The chief druidess begins her sermon, and as she talks about the rebirth of the sun from the womb of the mother goddess, I'm reminded of the womb-like chamber within the mound at Dowth.

Now, we each take a piece of shortbread, a dried apricot or a cornflake cake from the large platter being passed around and open our mouths to drink the rain – someone managed to leave the mead in the car park.

Modern druids don't have a fixed set of beliefs or practices that they adhere to, although nature is an important focus of their reverence, and many also believe that the soul is successively reincarnated, just as the sun is reborn each midwinter. They gather eight times during the yearly cycle in order to mark key turning points in our journey around the sun and the farming cycle it influences: the birth of lambs; the mating of livestock; the harvest; and the slaughter of animals in late autumn.

Two of our company step forward into the circle, one wearing a crown of oak, the other of holly, both brandishing wooden staves. They prod at one another, jeering abuse, until a fully blown fight breaks out. The crowd cheers them on, some heckling the 'oak-king' and some the 'holly-king', until the latter is pinned to the ground and asked to surrender. 'All right,' he mutters, as his holly crown rolls off into the sodden grass, 'but I'll get you next time'. He's referring to the next ritual battle, which will be re-enacted in six months' time at the summer solstice. This time, the holly-king will triumph, presiding over the autumn months as winter marches in.

As we soggily trudge back to our cars, I look up to the sky and spot a starling murmuration, a swooping mass of birds, wheeling and diving, performing a midwinter ritual of their own.

A few nights later, once the rain has cleared, I drive back out into the Wiltshire countryside to look at the night sky. Cranborne Chase, a mere stone's throw from Stonehenge, is among the darkest places in England, and currently bidding for dark-sky reserve status of its own. As I lie back on a blanket among the frosty blades of grass, and let my eyes adjust to these unfamiliar surroundings, I scour the skies for Orion's Belt, to get my bearings.

A friend once told me that in a single night it is possible to observe the life cycle of stars. Tracing the bright dots of Orion's sword, I spot the foggy cluster I'm looking for: the Orion Nebula, a nursery where new stars are being born.

From Orion, I also find the V-shaped face of Taurus the bull, and nearby the Pleiades, that frosty knot of stars that so inspired our ancestors.

There too is Betelgeuse, a star 600 times wider than our sun, which, were they sitting side by side, would be some 10,000 times brighter too. It is approaching the end of its stellar life; in the not too distant future, Betelgeuse will run out of fuel, collapse under its own weight, and then explode in a spectacular supernova. A similar fate awaits our own sun too, some day. Approximately 5 billion years from now, it will swell to such an enormous size that our planet, along with Mercury and Venus, will probably be engulfed by it – but then it will quietly fade away without becoming a supernova.

Betelgeuse is relatively close by in stellar terms, but when the photons from some of the more distant stars began hurtling through space towards us, humans – and in some cases, our planet – didn't even exist.

The next time you look up at the sun, or the stars, consider the effect of these photons on your biology as they're absorbed into your retina at the end of their epic journey. Light sparked life and has shaped our biology ever since, and it continues to influence us today. We are children of the sun, and we need its light as much as ever.

Acknowledgements

This book was in part inspired by my mother, Isobel Geddes, who for as long as I can remember, has divided the year along calendrical fault lines based on the availability of sunlight. Thanks to her for getting up in the blackness of midwinter and accompanying me to mark the sunrise at drizzly Stonehenge and Newgrange; for sharing her extensive knowledge about prehistoric monuments; and for being a valuable first reader.

This book couldn't have been written without the fantastic patience and parenting skills of my husband, Nic Fleming, who bravely held the fort while I ventured off on multiple trips to Scandinavia, the US, Germany and Italy; read my (very long) first draft and helped improve it; and provided encouragement during moments of literary despair. Thanks also to Nic and to our children, Matilda and Max, for humouring my 'darkness experiment' and spending several weeks during December and January living without any electric light.

I am extremely grateful to my agent, Karolina Sutton, and to Rebecca Gray at Profile Books for believing in my idea and commissioning me to write about it in the first place. Thanks too, to the Winston Churchill Memorial Trust, which funded so much of the travel that was necessary for this book, and to Mun-Keat Looi and Chrissie Giles at the Wellcome Trust's *Mosaic*, which also funded several of my overseas trips.

My visit to the Amish community in Lancaster County wouldn't have been possible without the help, trust and enthusiasm of Teodore Postolache at the University of Maryland, who introduced me to Hanna and Ben King. I'm grateful to Sonia Postolache for her companionship and excellent driving skills – and to Hanna and Ben for allowing me into their home, introducing me to their friends and family, and answering my relentless stream of questions about light, sleep and Amish life.

Neither would I have been able to spend the night on a Milanese psychiatric ward without the trust and assistance of Francesco Benedetti at San Raffaele Hospital in Milan. Grazie molto, also, to those patients who shared intimate details of their illnesses with me and to Irene Bollettini for acting as translator.

Richard Fisher at BBC Future gamely commissioned me to research the effects of living without electric light and provided the funding for some of the scientific tests. I am indebted to Derk-Jan Dijk and Nayantara Santhi at the University of Surrey who helped me to devise the experiment and analyse the data. Thanks also to Marijke Gordijn at Chrono@Work for carrying out the melatonin analysis, and to Frank Scheer at Brigham and Women's Hospital and Mariana Figueiro at the Lighting Research Center who also helped with the data interpretation.

As a science journalist, I am unceasingly indebted to the various researchers and individuals who spare the time to talk me through their work and experiences, and the same goes for this book: Even if you are not mentioned or quoted directly in these pages, the insights and explanations you have provided have been invaluable. Particular credit goes to Anna Wirz Justice, Derk-Jan Dijk and Prue Hart, for reading various chapters and providing feedback on the scientific accuracy of the material, and to my father-in-law, Andrew Fleming, for his thoughts on the archaeological content.

In researching the science of circadian rhythms, I leafed through countless journal articles and books, but found *Rhythms of Life* by Russell Foster and Leon Kreitzman; *Circadian Rhythms: a very short introduction* by the same authors; and *Sleep: a very short introduction* by Steven Lockley and Russell Foster great starting points. Also, highly recommended are *Internal Time* by Till Roenneberg and *Reset Your Inner Clock* by Michael Terman. I'm grateful to Professors Lockley, Terman and Roenneberg for taking the time to meet with me and answer my additional questions – particularly to Professor Lockley, who patiently taught me how to minimise jet lag, and may therefore have spared me this horror for the rest of my life. Matthew Walker's *Why We Sleep* was also extremely useful.

For my research into the effects of sunlight on our skin, I drew heavily on a themed collection of papers in *Photochemical and Photobiological Sciences: The health benefits of UV radiation exposure through vitamin D production and non-vitamin D pathways*. Richard Hobday's *The Healing Sun*, meanwhile, provides an excellent historical account of light therapy through the ages.

I spent a ridiculous amount of time researching our historical relationship with sunlight, most of which didn't make the final cut. However, for further reading about this fascinating subject, I'd highly recommend *Stations of the Sun* by Ronald Hutton, and *Prehistoric Belief* by Mike Williams. And for an excellent general overview of mankind's relationship with sunlight over the ages, *Chasing the Sun* by Richard Cohen is encyclopaedic in its breadth, while if you'd like to learn more about the evolution of electric light, Jane Brox's *Brilliant* lives up to its title.

Finally, thanks to the team at Profile and Wellcome Collection for helping to put together this book and market it – particularly to my editors Fran Barrie and Cecily Gayford, and to my copyeditor Susanne Hillen. There were many trees obscuring the wood, and together, you did an excellent job of clearing them.

Notes

Introduction

1. https://physoc.onlinelibrary.wiley.com/doi/full/10.1113/expphysio1.2012.071118.
2. Richard Cohen, *Chasing the Sun: The Epic Story of the Star That Gives Us Life* (Simon & Schuster, London, 2011), p. 292.
3. Q. Dong, 'Seasonal Changes and Seasonal Regimen in Hippocrates', *Journal of Cambridge Studies*, 6 (4), 2011, p. 128. https://doi.org/10.17863/CAM.1407.

1: The Body Clocks

1. At least, if studies in mice are to be believed. One recent study found that replication of the herpes virus was up to ten times greater if mice were infected at the start of their resting phase, compared to when they would usually be active; see http://www.pnas.org/content/early/2016/08/10/1601895113. Other studies have suggested that their vulnerability to food-borne pathogens is greater then as well; see https://www.cell.com/cell-host-microbe/pdf/S1931–3128(17)30290-1.pdf.
2. https://www.ncbi.nlm.nih.gov/pmc/articles/PMC3022154/.
3. Peter Coveney and Roger Highfield, *The Arrow of Time* (Penguin Books, London, 1990).
4. Benzer was inspired by the work of Colin S. Pittendrigh – widely considered the founding father of circadian rhythms. He was the first to show that *Drosophila* larvae emerge from their pupa like clockwork, even when kept in constant darkness.
5. A 'free-running period' is the scientific term for the amount of time it takes someone's endogenous or pre-programmed rhythm to repeat itself in the absence of environmental time cues such as light.

6. http://www.kentonline.co.uk/kent/news/ lifelong-islander-harry-loses-ca-a49624/.

7. Most blind people can still determine the difference between light and dark, and where a light source is coming from. Total blindness is the complete lack of light perception.

8. https://www.tandfonline.com/doi/abs/10.1080/ 00140138708966031.

9. http://www.tandfonline.com/doi/abs/10.3109/07420528.2016.11 38120. This study also found that the parents of 'evening-type' children reported more sleep-related challenges. Their kids were more likely to resist going to bed at night, to wake up in a negative mood, and to have conflicts with their parents.

10. David R. Samson et al., 'Chronotype variation drives night-time sentinel-like behaviour in hunter-gatherers', *Proceedings of the Royal Society B*, 284 (1858), 12 July 2017, doi: 10.1098/ rspb.2017.0967.

2: The Body Electric

1. For more on the evolution of electric light, see Jane Brox, *Brilliant* (Souvenir Press, London, 2011), which is a great read.

2. Robert Louis Stevenson, *Virginibus Puerisque*, 1881.

3. Robert Louis Stevenson, *Virginibus Puerisque*, 1881.

4. Jim Horne, *Sleepfaring: A Journey through the Science of Sleep* (Oxford University Press, 2007).

5. Nicholas Campion, in discussion with the author. For more on this idea, see Campion's preface to Ada Blair's *Sark in the Dark: Wellbeing and Community on the Dark Sky Island of Sark* (Sophia Centre Press, Bath, 2016), p. xvii.

6. https://www.scientificamerican.com/article/q-a-the-astronaut-who-captured-out-of-this-world-views-of-earth-slide-show1/.

7. To learn more about this project, visit http://citiesatnight.org/.

8. https://www.extension.purdue.edu/extmedia/fnr/fnr-faq-17. pdf.

9. https://www.nature.com/articles/nature23288.

10. https://www.ncbi.nlm.nih.gov/pmc/articles/PMC4863221/.

11. Donald J. Trump, *Think Like a Billionaire* (Ballantine Books, New York, 2005), p. xvii.

12. For a more detailed description of how sleep shaped our evolution, and the role it plays in memory and emotional regulation, see Matthew Walker's book, *Why We Sleep* (Allen Lane, London, 2017), pp. 72–77.

13. Russell Foster and Leon Kreitzman, *Circadian Rhythms: A Very Short Introduction* (Oxford University Press, 2017), p. 17.

14. http://www.cell.com/current-biology/abstract/S0960-9822(15)01157-4.

15. https://www.cell.com/current-biology/abstract/S0960-9822(13)00764-1.

16. http://www.cell.com/current-biology/fulltext/S0960-9822(16)31522-6.

17. For factories assembling electrical components, or other workplaces requiring perception of fine details, the HSE recommends an average illuminance of 500 lux.

18. http://www.sleephealthjournal.org/article/S2352-7218(17)30041-4/fulltext.

19. https://www.ncbi.nlm.nih.gov/pubmed/29040758.

20. https://www.ncbi.nlm.nih.gov/pubmed/28637029.

21. https://www.ncbi.nlm.nih.gov/pubmed/22001491.

22. http://www.sjweh.fi/show_abstract.php?abstract_id=1268.

23. http://www.sciencedirect.com/science/article/pii/S0165032712006982.

3: Shift Work

1. Arianna Huffington, *The Sleep Revolution: Transforming Your Life, One Night at a Time* (W.H. Allen, London, 2017).

2. Certainly, prolonged sleep deprivation appears deadly to rats: they die after approximately fifteen days of being kept awake – roughly as long as it takes for them to die without food. In the run-up to death, they lose their ability to regulate body temperature, develop wounds and sores on their skin and internal organs, and their immune system collapses.

3. https://www.ncbi.nlm.nih.gov/pmc/articles/PMC1739867/.

4. *Acute Sleep Deprivation and Risk of Motor Vehicle Crash Involvement* (AAA Foundation for Traffic Safety, December 2016).
5. https://www.ncbi.nlm.nih.gov/pmc/articles/PMC4030107/.
6. Foster and Kreitzman, *Circadian Rhythms*, p. 19.
7. Study presented at the annual meeting of the Associated Professional Sleep Societies in Boston in June 2017, by Sierra B. Forbush of the University of Arizona.
8. Till Roenneberg, in discussion with the author.
9. Seth Burton, in discussion with the author.
10. https://www.ncbi.nlm.nih.gov/pubmed/10704520.
11. https://journals.plos.org/plosone/article?id=10.1371/journal.pone.0015267.
12. http://www.pnas.org/content/115/30/7825.
13. Richard Stevens, in discussion with the author.
14. https://www.ncbi.nlm.nih.gov/pubmed/8740732.
15. http://www.pnas.org/content/pnas/106/11/4453.full.pdf.
16. https://www.ncbi.nlm.nih.gov/pubmed/26548599.
17. https://onlinelibrary.wiley.com/doi/abs/10.1002/oby.20460.
18. Jonathan Johnston, in discussion with the author.
19. http://www.cell.com/current-biology/abstract/S0960-9822(17)30504-3.
20. https://www.sciencedaily.com/releases/2017/08/170815141712.htm.
21. https://www.ncbi.nlm.nih.gov/pubmed/22621361.

4: Doctor Sunshine
1. For further reading on the fascinating history of 'sun cures', I'd highly recommend Richard Hobday's *The Healing Sun: Sunlight and Health in the 21st Century* (Findhorn Press, Forres, 1999).
2. Florence Nightingale, *Notes on Nursing: What it is, and What it is not* (CreateSpace Independent Publishing Platform, 2015).
3. https://www.ncbi.nlm.nih.gov/pubmed/15888127.
4. Hobday, *The Healing Sun*.
5. https://www.ncbi.nlm.nih.gov/pmc/articles/PMC3277100/.

6. Quoted in Joseph Mercola, *Dark Deception: Discover the Truths about the Benefits of Sunlight Exposure* (Thomas Nelson, Nashville, Tennessee, 2008).
7. http://www.jbc.org/content/64/1/181.full.pdf.
8. From Paul Jarrett and Robert Scragg, 'A short history of phototherapy, vitamin D and skin disease', *Photochemical & Photobiological Sciences*, vol. 3, 2017.
9. Victor Dane, *The Sunlight Cure: How to Use the Ultraviolet Rays* (Athletic Publications, London, 1929).
10. Quoted in Jarrett and Scragg, 'A short history of phototherapy, vitamin D and skin disease', 2017.
11. Quoted in Hobday, *The Healing Sun*.
12. https://www.thelancet.com/journals/lancet/article/PIIS0140-6736(16)31588-4/fulltext.
13. https://www.telegraph.co.uk/news/2018/01/12/no-light-end-tunnel-chelseas-new-1-billion-stadium/.
14. https://www.hindustantimes.com/delhi-news/in-a-dense-and-rising-delhi-exert-your-right-to-sunlight/story-zs0xLKVT8UKC05B5JfQi5M.html.
15. https://www.aaojournal.org/article/S0161-6420(07)01364-4/fulltext.
16. Ian Morgan, in discussion with the author.
17. https://www.ncbi.nlm.nih.gov/pubmed/26372583.

5: Protection Factor

1. https://www.ncbi.nlm.nih.gov/pubmed/2003996.
2. https://www.newscientist.com/article/mg19325881-700-born-under-a-bad-sign/.
3. https://www.ncbi.nlm.nih.gov/pmc/articles/PMC4986668/.
4. http://journals.plos.org/plosbiology/article?id=10.1371/journal.pbio.1000316.
5. Besides multiple sclerosis, one of the most robust associations for these month-of-birth effects is for type 1 diabetes – another autoimmune disease; see https://www.ncbi.nlm.nih.gov/pmc/articles/PMC2768213/.
6. These figures only apply to nations containing people of

primarily European descent. For other nations, no association with latitude was found – but Europeans have a higher genetic risk of MS in the first place. See https://www.ncbi.nlm.nih.gov/pubmed/21478203.

7. Consistent annual data wasn't available before this point.

8. https://www.karger.com/Article/Abstract/336234.

9. https://www.karger.com/Article/FullText/357731.

10. https://www.sciencedirect.com/science/article/pii/B9780128099650000331.

11. https://www.ncbi.nlm.nih.gov/pmc/articles/PMC4861670/.

12. https://www.newscientist.com/article/mg22329810-500-let-the-sunshine-in-we-need-vitamin-d-more-than-ever/.

13. http://journals.sagepub.com/doi/full/10.1177/1352458517738131.

14. https://www.ncbi.nlm.nih.gov/pubmed/29102433.

15. https://www.ncbi.nlm.nih.gov/pubmed/4139281/.

16. Scott Byrne, in discussion with the author.

17. https://www.omicsonline.org/open-access/uv-irradiation-of-skin-regulates-a-murine-model-of-multiple-sclerosis-2376-0389-1000144.php?aid=53832.

18. https://www.ncbi.nlm.nih.gov/pmc/articles/PMC5954316/.

19. Richard Weller, in discussion with the author.

20. In one study, Weller and his colleagues exposed people to 22 minutes of UVA light, and recorded a drop in diastolic blood pressure that was maintained for 30 minutes after the light was switched off: see https://www.ncbi.nlm.nih.gov/pubmed/24445737.

21. https://www.ncbi.nlm.nih.gov/pubmed/25342734.

22. https://www.ncbi.nlm.nih.gov/pubmed/26992108.

23. Even sun avoiders can get melanoma – possibly because of sunburn during childhood.

24. Another rich source of vitamin D is oily fish, which provides many other nutrients besides.

6: A Dark Place

1. http://www.rug.nl/research/portal/files/3065971/c2.pdf.

2. From p. 5 of these excerpts: http://www.five-element.com/graphics/neijing.pdf.

3. Quoted in Russell Foster and Leon Kreitzman, *Seasons of Life* (Profile Books, London, 2009), p. 200–201.

4. Foster and Kreitzman, *Seasons of Life*.

5. According to the *Diagnostic and Statistical Manual of Mental Health Disorders*, Fifth Edition (DSM-5), which is widely used by psychiatrists, seasonal affective disorder is a subtype form of depression – major depressive disorder with a seasonal pattern. To receive a diagnosis, patients must therefore meet the diagnostic criteria for recurrent major depression or bipolar mood disorder – the difference is that their symptoms display a seasonal pattern; see https://bestpractice.bmj.com/topics/en-gb/985.

6. For a comprehensive history of seasonal affective disorder, I'd recommend C. Overy and E. M. Tansey, eds, *The Recent History of Seasonal Affective Disorder (SAD)*, the transcript of a Witness Seminar held by the History of Modern Biomedicine Research Group, Queen Mary, University of London, on 10 December 2013; see http://www.histmodbiomed.org/sites/default/files/W51_LoRes.pdf.

7. https://jamanetwork.com/journals/jamapsychiatry/article-abstract/494864.

8. https://www.ncbi.nlm.nih.gov/pubmed/6581756.

9. https://www.ncbi.nlm.nih.gov/pubmed/2326393.

10. https://www.ncbi.nlm.nih.gov/pmc/articles/PMC4673349/.

11. https://www.arctic-council.org/index.php/en/about-us/member-states/norway.

12. https://www.ncbi.nlm.nih.gov/pubmed/8250679.

13. http://journals.sagepub.com/doi/10.1177/070674370204700205.

14. Overy and Tansey, eds, *The Recent History of Seasonal Affective Disorder (SAD)*, 2013.

15. https://theconversation.com/a-small-norwegian-city-might-hold-the-answer-to-beating-the-winter-blues-51852.

16. Kari Leibowitz, in discussion with the author.

17. Overall, CBT and light therapy appear comparable in terms of

reducing SAD symptoms, but certain symptoms (trouble falling asleep, excessive sleepiness, anxiety and social withdrawal) reduced more quickly in response to light therapy; see https://www.ncbi.nlm.nih.gov/pubmed/29659120.
18. https://www.ncbi.nlm.nih.gov/pubmed/26539881.

7: Midnight Sun
1. http://www.pbs.org/wgbh/nova/earth/krakauer-in-antarctica.html.
2. Foster and Kreitzman, *Seasons of Life*, p. 221.
3. They block the reuptake of serotonin, meaning that it sticks around in the junctions between neurons for longer, and therefore has more of an effect.
4. 'Sex differences in light sensitivity impact on brightness perception, vigilant attention and sleep in humans', S. L. Chellappa et al, in *Scientific Reports* 7, article no. 14215 (2017). Also 'Influence of eye colours of Caucasians and Asians and the suppression of melatonin secretion by light', S. Hrguchi et al, in *American Journal of Physics – Regulatory, Integrative and Comparative Physiology*, Vol. 292, issue 6.
5. Most studies that have investigated this have done so in the context of a hospital-like environment, where the lights often remain switched on 24/7, and noise can similarly disrupt sleep.
6. It is still not perfect: the light from current dawn simulation clocks is far dimmer than daylight, and such devices are usually positioned behind people's heads, rather than in front of them, meaning that less light reaches their eyes.
7. Hot baths before bed have been found to boost NREM sleep by 15 to 20 per cent; see Walker, *Why We Sleep*, p. 279.

8: Light Cure
1. Her name has been changed to protect her identity.
2. https://journals.plos.org/plosone/article?id=10.1371/journal.pone.0033292.
3. https://www.ncbi.nlm.nih.gov/pubmedhealth/PMH0021785/.
4. A recent study at the ICU at Central Manchester Foundation

Trust found an average daytime illuminance of 159 lux, which
is 10 to 1,000 times dimmer than daylight, while night-time
illuminance averaged 10 lux – around fifty times brighter than
moonlight. Overnight, there were also several bright pulses
of light (measuring up to 300 lux), because of procedures or
checks being carried out.

5. https://www.ncbi.nlm.nih.gov/pmc/articles/PMC4507165/.
6. http://journals.sagepub.com/doi/
   full/10.1177/1477153512455940.
7. https://www.ncbi.nlm.nih.gov/pmc/articles/
   PMC1296806/?page=2.
8. https://www.ncbi.nlm.nih.gov/pubmed/27733386.
9. https://www.thelancet.com/journals/lancet/article/
   PIIS0140–6736%2817%2932132–3/fulltext?elsca1=tlpr.
10. http://stm.sciencemag.org/content/9/415/eaa12774.
11. http://stm.sciencemag.org/content/9/415/eaa12774.
12. http://www.cochrane.org/CD006982/NEONATAL_cycled-
    light-intensive-care-unit-preterm-and-low-birth-weight-infants.
13. Data presented at the 2017 meeting of the Society for Light
    Therapy and Biological Rhythms in Berlin.
14. https://jamanetwork.com/journals/jama/fullarticle/273623.
15. http://www.pnas.org/content/111/45/16219.
16. https://www.ncbi.nlm.nih.gov/pubmed/4076288.
17. https://www.ncbi.nlm.nih.gov/pubmed/2179481.
18. https://www.ncbi.nlm.nih.gov/pubmed/22745214.
19. https://www.ncbi.nlm.nih.gov/pubmed/22745214.
20. https://www.ncbi.nlm.nih.gov/pmc/articles/PMC4874947/.
21. The full quote is: 'It is often thought that medicine is the
    curative process. It is no such thing; medicine is the surgery
    of functions, as surgery proper is that of limbs and organs.
    Neither can do anything but remove obstructions; neither
    can cure; nature alone cures. Surgery removes the bullet out
    of the limb, which is an obstruction to cure, but nature heals
    the wound. So it is with medicine; the function of an organ
    becomes obstructed; medicine so far as we know, assists nature
    to remove the obstruction, but does nothing more. And what

nursing has to do in either case, is to put the patient in the best condition for nature to act upon him.' *Florence Nightingale: The Nightingale School*, Lynn McDonald, ed., Wilfrid Laurier University Press, Waterloo, Ontario, 2009, p. 683.

9: Fine-tuning the Clock

1. http://www.espn.co.uk/nba/story/_/id/17790282/the-nba-grueling-schedule-cause-loss.
2. Research presented at the Sleep 2017 meeting in Boston.
3. 'The effects of sleep extension on the athletic performance of collegiate basketball players', C. D. Mah et al, *Sleep* 2011; 34(7): 943–50.
4. Kevin Bickner, in an interview with Ben Cohen, *Wall Street Journal*, 7 February 2018.

10: Clocks for Society

1. Kantermann describes this in the TED talk he gave in Groningen 2016.
2. https://www.theatlantic.com/health/archive/2014/02/the-town-thats-building-life-around-sleep/283553/.
3. There's less call for it in countries closer to the equator, where dawn and dusk are more uniform throughout the year.
4. For early-rising farmers, at some latitudes, DST robs them of morning daylight.
5. Winston Churchill, 'A Silent Toast to William Willett', *Finest Hour* (Journal of the International Churchill Society), 114, spring 2002.
6. https://www.bmj.com/content/1/2632/1386.
7. https://www.ncbi.nlm.nih.gov/pmc/articles/PMC4513265/.
8. http://psycnet.apa.org/record/2010-22968-001.
9. https://www.ncbi.nlm.nih.gov/pubmed/22369272.
10. https://www.bbc.co.uk/news/world-europe-45366390.
11. According to plans put forward in a Private Member's Bill in 2010, the UK would still retain DST, so we would in effect have double summer time between late March and late October.

12. https://www.cell.com/current-biology/pdf/S0960–9822(06)02609–1.pdf.

13. Here, the difference seems to be 2 minutes per degree – although possibly the data is less accurate because the US population is so concentrated in urban areas – making the difference between an eastern state like Maine and a western one like Indiana – both of which are within the Eastern time zone – approximately 40 minutes. I wrote about this research in *New Scientist*: https://www.newscientist.com/article/2133761-late-nights-and-lie-ins-at-the-weekend-are-bad-for-your-health/.

14. The original reason for the switch was a Russian Academy of Medical Sciences report, which stated that, when the clocks changed, there was a 1.5-fold increase in heart attacks, and that the rate of suicides grew by 66 per cent.

15. Social jet lag is defined as the difference between your midpoint of sleep on work (or school) days, and the midpoint on free days. Say you go to bed at 11 p.m. and wake at 7 a.m. on work days (mid-sleep at 3 a.m.) and go to bed at 2 a.m. and wake at 10 a.m. on weekends (mid-sleep at 6 a.m.), you would experience three hours of social jet lag per week.

16. https://www.sciencedirect.com/science/article/pii/S0960982212003259.

17. Teenagers need more sleep than adults, so it's important to let them catch up on missed sleep rather than turfing them out of bed on a Saturday morning. It's far better to encourage them to go to bed earlier all week-round, by maximising daylight exposure, and by minimising exposure to blue light in the evenings.

18. https://www.sciencedirect.com/science/article/pii/S0262407917317700.

19. https://www.ncbi.nlm.nih.gov/books/NBK222802/.

20. https://conservancy.umn.edu/bitstream/handle/11299/4221/CAREI%20SST-1998VI.pdf?sequence=1&isAllowed=y.

21. https://www.ncbi.nlm.nih.gov/pubmed/20603459.

22. http://pediatrics.aappublications.org/content/pediatrics/early/2014/08/19/peds.2014–1697.full.pdf.
23. https://www.frontiersin.org/articles/10.3389/fnhum.2017.00588/full.
24. Foster and Kreitzman, *Circadian Rhythms*, p. 15.
25. http://mikechristian.web.unc.edu/files/2016/11/Christian-Ellis-SD-AMJ-2011.pdf.
26. http://journals.sagepub.com/doi/abs/10.1177/0956797614541989?journalCode=pssa
27. https://www.ncbi.nlm.nih.gov/pubmed/29073416.

Index

free radicals *see* reactive oxygen
  species
fruit fly experiments 23–6, 37
Fukushima Daiichi 85

G
Galvin, Mark 16–18, 22, 32–6,
  38
gambling 2–3, 185
gas lighting 42–3
genes
  under circadian control 13
  CRY1 37
  'period' and 'timeless' 25
genetics
  and chronotypes 36
  'clock genes' 18, 30, 162
  and myopia 106
  and SAD 140
Germany 26, 63–4, 158, 189–
  91, 193, 195, 203
Gordijn, Marijke 59, 63
grandparents, utility of 38
green spaces, urban 102–5
Greenacre Park, NY 102–4
GSA (US General Services
  Administration) 62, 65
Guangzhou, China 105, 108

H
Hadza people, Tanzania 37–8,
  53
Hallkvist, Jan 85–7
Hanna (King, Amish
  community member) 40,
  51–2, 64

Harris, Tony 104
Hart, Prue 119
heart attacks 164, 193
Hess, Alfred 98
Hilgers, Hubertus 193
Hippocrates 7–8
homeostatic regulation 28–9
hormones 10–11, 68, 75, 79–80,
  132, 166
  see also adrenaline; cortisol;
    insulin; melatonin
hospital lighting 92, 100, 162–
  4, 166–7
hypothalamus: 19, 132

I
Iceland 140
illnesses
  disease susceptibility and
    birth date 111–14s
  influenced by circadian
    rhythms 13, 25–6, 69,
    71–2, 131–2, 158, 162
  metabolic syndrome 81–2
  see also mental illness
illuminance levels 53–5, 60–1,
  87, 163, 169, 178
immune system
  circadian disruption 162
  recovery from heart attacks
    164
  suppression by UV 118–19
India 6, 104, 172
indoor lighting
  hospital lighting 92, 100,
    162–4, 166–7

illuminance levels 54–5,
    60–1, 87, 163, 169
office lighting 12, 55, 61, 64
schools 64
tailoring 204
infants
    premature and low birth-
        weight 165–6
    prenatal sunlight exposure
        11, 112
    prenatal vitamin D effects
        109
    sleep patterns 10
infrared, therapeutic use 7
insomnia 9, 53, 130, 150, 196
insulin 68, 74, 79, 81–2, 113
International Agency for
    Research on Cancer 12,
    76–7
International Space Station
    (ISS) 47, 177–9, 188
ipRGCs (intrinsically
    photosensitive retinal
    ganglion cells) 29–30, 65
Iran 113–14, 125
isolation experiments 26–7
ISS (International Space
    Station) 47, 177–9, 188

J
jaundice 9
jet lag
    adjusting to/recovering from
        14, 31, 175–6, 180–3
    in astronauts 179–80
    in athletes 183–7

in bees 22–3
causes and effects 9–10
effect of travel direction 181,
    185
social jet lag 69, 192, 194–5,
    197
in submariners 71
see also daylight saving; time
    zones
Johnston, Jonathan 81–2
Johnston, Smith L 178–80

K
Kantermann, Thomas 190–1,
    194
Kennett, Harry 28, 31
Kern, Herb 128–9, 136
kidneys 10, 74, 172
King, Hanna and Ben 40, 51–2,
    64
Kittilsen, Oscar 135–6
Koch, Robert 92
Konopka, Ronald 24
Krakauer, Jon 149
Kripke, Daniel 132
Kripke, Margaret 118
Kunz, Dieter 64

L
'larks'
    amongst the Amish 52–5
    early risers as 18, 35, 37, 150,
        195–6, 200–1
    eastward travellers becoming
        180, 182
Las Vegas 2, 11–12